OFFSHORE TECHNOLOGY IN CIVIL ENGINEERING

Hall of Fame Papers

VOLUME 12

SPONSORED BY
Program Committee, Offshore Technology Conference of the
Coasts, Oceans, Ports, and Rivers Institute of the
American Society of Civil Engineers

EDITED BY

J. S. Templeton III, Dr.Eng., P.E.
Alan G Young, P. E.

ASCE

COPRI
COASTS, OCEANS, PORTS & RIVERS INSTITUTE

Published by the American Society of Civil Engineers

Library of Congress Cataloging-in-Publication Data

Offshore technology in civil engineering: Hall of Fame papers from the early years / sponsored by Subcommittee on Offshore Technology Conference Support, Coasts, Oceans, Ports, and Rivers Institute (COPRI) of the American Society of Civil Engineers; edited by J. S. Templeton III.

p. cm.

Includes bibliographical references and index.

ISBN 0-7844-0871-8

1. Ocean engineering—Congresses. 2. Offshore structures—Congresses. I. Templeton, J. S. (Jack S.) II. Coasts, Oceans, Ports and Rivers Institute (American Society of Civil Engineers). Subcommittee on Offshore Technology Conference Support.

TC1505.O37 2006
627'.98—dc22

2006042804

Published by American Society of Civil Engineers
1801 Alexander Bell Drive
Reston, Virginia, 20191-4382
www.asce.org/bookstore | ascelibrary.org

ISBN 978-0-7844-1478-1 (print)
Manufactured in the United States of America.

22 21 20 19 18 17 1 2 3 4 5

Preface

This is the twelfth volume in a series of publications containing classic papers from the Proceedings of the Offshore Technology Conference (OTC), the world's leading event for the development of offshore resources in the fields of exploration, drilling, production, and environmental protection. The American Society of Civil Engineers (ASCE), through its participation in and support of the OTC, plays a major role in the innovation and evolution of the technologies needed to overcome the challenges facing development of resources in the offshore environment. The years since the first OTC Conference in 1969 have seen the presentation of over 10,000 papers in the various technical disciplines central to offshore development. Several of these papers, presented throughout OTC's history, provided innovation in, vision for and lasting impact on the design, construction, and installation of the offshore infrastructure. Many have been adopted by design standards worldwide. Many have become integral methods incorporated in widely used design software. Some have had influence far beyond the offshore industry, and some have become integral to the design process of other structures such as buildings and bridges.

The ASCE OTC Committee has been at the forefront in recognizing the value the civil engineering technical papers provide to the engineering community and industry in general. In 2004, the committee began presenting ASCE awards for the OTC's Best Paper of the Year in Civil Engineering. In 2005 the committee decided to elevate and expand this recognition to include outstanding papers from the entire history of OTC, through the creation of a Hall of Fame for OTC Papers in the Civil Engineering Arena. A special sub-committee was charged with developing stringent criteria for nomination and selection of the Hall of Fame papers. Calls for nomination were made, the nominated papers were evaluated, and the first papers were selected for award and induction into the Hall of Fame in 2006. The process was repeated in 2007 and has now become an annual undertaking. In this, its twelfth year, the Hall of Fame once again recognizes a group of classic offshore technology papers whose significance has withstood the test of time.

This volume is a collection of the five winning papers inducted in 2017 at an award ceremony during OTC in May of 2017. The engineering methods published in these papers have proven their value through widespread use, permeating codes, standards, guidelines, and engineering software. Even though these papers were presented at OTC conferences more than twenty years ago, they still form the core of current practice worldwide.

Acknowledgments

The Editors and the ASCE OTC Program Committee gratefully acknowledge OTC, COPRI, and ASCE for their support in making the publication of this commemorative volume possible, and to ASCE for providing the seed money to fund the Gala Dinner honoring the awardees. The Editors would also like to thank members of the ASCE OTC Program Committee of Offshore Technology Conference, for their support in selecting the papers worthy to receive the Hall of Fame Award.

ASCE OTC Program Committee Members

James S. Chitwood, P.E., *Chair*
Sathish Ramamoorthy, *Vice Chair*

Puneet Agarwal, Ph.D., P.E.
David Barton, P.E.
Eugene P. Berek
Subir Bhattacharjee, Ph.D.
Ian C. Chu, P.E.
James E. Dailey, Ph.D., P.E.
Ron de Jong
Matthew Ferrell
Chris Hadley, P.E.
Randall Hance
Patrick J. Hudson, Ph.D., P.E.
Philippe Jeanjean, Ph.D.
Vikrant V. Joshi
Sai Majhi
Edward Mascorro, P.E.
Thomas M. Miller
Jason A. Newlin, P.E.
Daniel R. Spikula, P.E.
John S. Templeton III, Dr. Eng., P.E
David J. Wisch, P.E.
Alan G. Young, P.E.

In the process of nominating, evaluating, and recommending papers for the awards, a Hall of Fame Panel composed of past recipients of the Hall of Fame awards was selected. The panel submitted nominations and solicited others from the ASCE OTC Program Committee and other industry representatives. The panel met many times to review the nominations and to select the award-winning papers.

The members of this panel were selected to represent a broad range of technical disciplines. The panel selected Mr. Alan Young as their chairman. The panel provided insight and judgment of great value in the nomination and selection process. We hereby acknowledge their significant contribution, and thank them for the generous time they devoted to the nomination, evaluation and selection process.

Hall of Fame Panel Members

Alan G Young, P.E., *Chair*

Gene Berek
Brad Campbell, Ph.D
Earl Doyle, P.E. S.E.
David B. Driver
Peter Marshall, P.E.
J. D. Murff, P.E.
Gordon Sterling, P.E.
J.S. Templeton, III, Dr. Eng., P.E.
E.G. Ward, P.E.

Dedication

This volume is dedicated to all of the authors who have contributed to the success of Offshore Technology Conference throughout its long history. Their personal and professional commitment and devotion to document and share their new ideas for concepts, design, construction, installation, and research, through presentations and dissemination in the OTC proceedings, has been a key in accelerating the development of the strategic energy industry.

Finally, this dedication honors the memory of all deceased OTC authors, especially authors of papers included in this volume. We mark their passing with respect for their leadership and with gratitude for their important contributions to the offshore industry.

OFFSHORE TECHNOLOGY IN CIVIL ENGINEERING

Hall of Fame Award Papers

2006 Award Papers — Volume One

Paper Number 1006 (1969) J.D. Wheeler

Method for Calculating Forces Produced by Irregular Waves

The paper described and evaluated a new method for calculating wave forces on offshore structures. It accounted for linear filtering of surface profile records into the frequency domain, superposition of Airy wavelets in the time domain, and stretching of crest kinematics, all essential elements of nonlinear dynamic analysis of deepwater structures. The accuracy of the method has been verified by comparison with measured hurricane-wave forces. Wheeler stretching, as it became known, is common practice today.

Paper Number 1008 (1969) P.M. Aagaard and R. G. Dean

Wave Forces: Data Analysis and Engineering Calculation Method

This paper presented a new method for calculating ocean wave forces on offshore structures. It was based on data from two full-scale wave force measurement installations in the ocean and a mathematical model representing Morrison's hydrodynamic forces on submerged bodies in unsteady flow and Dean's kinematic flow field of highly nonlinear waves. The method was verified, and forms the basis for offshore structure design in RP2A and all major analysis software.

Paper Number 1043 (1969) R. M. Carter, P. W. Marshall, P.D. Thomas and T.M. Swanson

Materials Problems in Offshore Platforms

This paper presented failure investigations and other research leading to new criteria for tubular joint design and material selection relative to static strength (punching shear), fatigue resistance (hot spot strain), weldability (carbon equivalent), homogeneity (lamellar tearing), and economy. It is still the basis for AWS design criteria and the API steel specs. Although ISO-API strength equations now appear in a different format, some of them are still pure punching shear

Paper Number 1055 (1969) Carl H. Bender, Jr., Charles G. Lyons and Lee L. Lowery, Jr.

Applications of Wave-Equation Analysis to Offshore Pile Foundations

This paper presented a wave equation solution for pile driving and installation analysis and design. The method used impact stresses in both pile and hammer to optimize pile size, hammer, driving accessories, and field procedures. It illustrated practical applications to design and installation of heavily loaded piles. It was verified with extensive calibration in different soil types. Its principles are widely used to perform driveability studies and fatigue analysis for single-pile tendon anchorages.

Paper Number 1073 (1969) Carl G. Langner

The Articulated Stinger: A New Tool for Laying Offshore Pipelines

This paper presented a new method, referred to as tension S-laying, for offshore pipeline installation. It used a stinger with rollers as support structure, consisting several segments of uniform length, each segment being adjustably buoyant. Segments are connected in series by special hinge joints which provide a limited degree of vertical, lateral, and torsional flexibility. Stinger vertical flexibility increases water depth capability, and lateral flexibility increases weather capability. It is the method of choice in moderate water depths. Its analytical tools also apply to analysis of J-lay and catenary risers in deeper water.

Paper Number 1204 (1970) Hudson Matlock

Correlations for Design of Laterally Loaded Piles in Soft Clay

This paper first introduced the theory and framework for a fundamentally and experimentally based method of evaluating lateral soil reactions for pile foundations, using beam-column analysis. It further detailed the particulars of the method for use in soft clay soils. Its application was further extended to sand and stiff clay. It has been adopted worldwide, both within and outside the offshore industry, and is currently incorporated in most applicable codes and practices. It provided a method for successful pile design, averting failures using earlier simplified methods.

Paper Number 2080 (1974) Lymon C. Reese, William R. Cox and Francis D. Koop.

Analysis of Laterally Loaded Piles in Sand

This paper presented families of curves showing soil resistance p as a function of pile deflection y in fine sand or silty fine sand. These were based on soil behavior theory and results of experimental data for two large driven piles, instrumented for measuring bending moments, and subjected to short-term static and cyclic loading. Curve predictions compared favorably with actual behavior. The procedures used in developing these are curves form the design basis for API and other standards and codes.

Paper Number 2312 (1975) Lymon C. Reese, William R. Cox and Francis D. Koop.

Field Testing and Analysis of Laterally Loaded Piles in Stiff Clay

This paper presented families of curves showing soil resistance p as a function of pile deflection y in stiff clays. These were based on soil behavior theory and results of experimental data for two large and four small piles, instrumented for measuring bending moments, driven into stiff clay and subjected to short-term static and cyclic loading. The small piles were also pulled, redriven, and reloaded. Curve predictions compared favorably with actual behavior. The procedures used in developing these are curves form the basis of API and other standards and codes worldwide.

Paper Number 2537 (1975) Joseph Kallaby and David Millman

Inelastic Analysis of Fixed Offshore Platforms for Earthquake Loading

The paper introduced a method, commonly referred to as "pushover", which brought to the forefront the important role that ductility and reserve strength play in the design and assessment of platforms, or other structures, under extreme or "ductility level" earthquake events. It formed the basis for API RP 2A's revolutionary new earthquake criteria, adopted in 1975, in establishing platform ductility and reserve strength. Over a decade and a half later, it was adopted by API RP2A to evaluate reserve strength under storm conditions. It is now used worldwide, and is incorporated in all major analysis software. It has also become a standard structural engineering practice for performance-based evaluation of buildings.

2007 Award Papers — Volume Two

Paper 1590 (1972) H. O. Jahns and J. D. Wheeler

Long-Term Wave Probabilities Based on Hindcasting of Severe Storms

This paper established procedures to estimate the statistical distributions of wave parameters in severe storms such as hurricanes. It established the concept that wave crest height and trough depth may be equally important as wave height, or more so, in certain situations. Methods presented in this paper will be used to re-examine wave conditions experienced in recent severe hurricanes. Presently accepted design criteria are likely to be revised as a result.

Paper 2205 (1975) J. G. Kuang, A. B. Potvin and R. D. Leick
Stress Concentration in Tubular Joints.
This paper developed the first practical methodology for determination of tubular joint stress concentration factors. Thus it enabled routine fatigue assessment of platform joints. Prior to this, estimation of stress concentration factors for tubular joint design was impractical. The approach presented in this paper has been embodied in recommended practice documents since the late 1970's and is still routinely used for platform designs worldwide.

Paper 2370 (1975) L. M. Kraft, Jr., and J. D. Murff
A Probabilistic Investigation of Foundation Design for Offshore Gravity Structures
This paper presents a procedure for probabilistic analysis of offshore gravity structure foundations. It is one of the first examples of reliability based foundation design. It provided a new framework for analyzing uncertainty in design as well as probability of survival of the system. It pioneered load and resistance factor design for offshore foundations. This and the first order, second moment, uncertainty analysis method are used to the present day.

Paper 2371 (1975) A. G Young, L. M. Kraft, Jr., and J. A. Focht, Jr,
Geotechnical Considerations in Foundation Design of Offshore Gravity Structures
Prior to existence of any substantial safe operating history for offshore gravity structures, this paper set out the methodology for their complete foundation design including bearing capacity, sliding resistance, skirt penetration, foundation movements, and stability as well as site investigation. The methods given in this paper continue to be used today not only for gravity structures but also for mud mats, skirts, and other shallow foundations.

Paper 2608 (1976) J. H. Vughts and R. K. Kinra
Probabilistic Fatigue Analysis of Fixed Structures
This paper was the first to detail methodology to perform member-by-member fatigue analysis for offshore structures. It showed the feasibility of spectral analysis methods to predict fatigue performance, and it introduced the notion of a probabilistic approach to describe the environment and resulting stress. This set the direction that the industry took. The methodologies described in this paper are still in use to day, both for fixed and floating structures.

Paper 2688 (1976) L. D. Finn
A New Deepwater Offshore Platform – The Guyed Tower
This paper introduced one of the most innovative and important concepts in the history of offshore technology. The guyed tower, itself, was a major factor the offshore industry's ability to step off of the continental shelf and into truly deep water. Moreover, the concept of a compliant structure introduced here enabled a succession of other deep water compliant structures and also significantly continues to influence various floating production systems designs.

Paper 2896 (1977) J. D. Murff and T. W. Miller

Stability of Offshore Gravity Structures by the Upper Bound Method

This paper provided the first rigorous yet broadly applicable method for determination of foundation capacity for large offshore gravity base structures. This work, cited by API for shallow foundations design, not only has had a major impact on understanding the stability of conventional offshore gravity foundations, but it has also been adapted to analysis of multi-footing foundations, piles, caissons, slopes and mining works.

Paper 3230 (1978) R. E. Haring and J. C. Heideman

Gulf of Mexico Rare Wave Return Periods

The paper presented two new, innovative, ways of statistically analyzing hurricane hindcast data: pooling of data as a means of accounting for track variability and a kinematics reduction factor method for wave force calculation. Both of these became standard offshore practice. This paper has been widely cited in codes, publications and reports, and it is still influencing the development of new hurricane criteria accounting for intense, recent storms.

Paper 3777 (1980) L. P. Krolikowski and T. A. Gay

An Improved Linearization Technique for Frequency Domain Riser Analysis

This paper introduced advances in frequency domain analysis for riser evaluation that significantly improved modeling of riser fatigue damage for combination of sea states and vessel motions. The linearization techniques introduced in this paper have been broadly cited and are still in widespread use for risers and other offshore systems.

2008 Award Papers — Volume Three

Paper 1190 (1970) C. Petrauskas and P. M. Aagard

Extrapolation of Historical Storm Data for Estimating Design Wave Heights

This is the first OTC paper addressing the statistical techniques used in establishing design criteria. The principles and methods found in the paper were used widely and were considered as providing a better method for developing design criteria. The work is cited in most published papers pertaining to the estimation of wave design criteria.

Paper 1261 (1970) K. A. Blenkarn

Measurement and Analysis of Ice Forces on Cook Inlet Structures

This paper clearly demonstrated the value of field measurements in reducing the large uncertainties that are associated with ice loads on offshore platforms. The paper was written following the installation of 14 offshore platforms in Cook Inlet at a time of large uncertainties regarding interaction of sea ice with platforms. This information continues to be of current interest today for platforms in harsher arctic environments

Paper 1327 (1971) E. H. Doyle, B. McClelland, and G. H. Ferguson

Wire-Line Vane Probe for Deep Penetration Measurements of Ocean Bottom Sediment Strength

This paper introduced the use of the remote vane for in-situ strength testing of sub-marine soils, the first effective means for offshore in-situ testing of cohesive soils. It resulted in the discovery that conventional site investigation practice was introducing significant sample disturbance. More importantly this paper literally initiated modern in situ testing and undisturbed sampling from a floating platform. The concepts and principles laid out in this paper are integral components of all site investigations carried out today, especially those in deep water..

Paper 1500 (1971) G.F.M.Remery and A.J.Hermans

The Slow Drift Oscillations of a Moored Object in Random Seas

This paper developed a method to calculate the resonance motions of the moored ship and its mooring loads excited by second order wave effects. It addressed a physical problem of low frequency response that was not understood at the time. The discoveries made in this paper impacted the further development of the calculation codes to determine the wave drift loads with potential diffraction theory. The methodology was incorporated into analysis codes and the widely cited theory is still used today in potential flow diffraction codes.

Paper 2942 (1977) R. C. Kirby and C. P. Wroth

Application of Critical State Soil Mechanics to the Prediction of Axial Capacity for Driven Piles in Clay

The concepts presented in this paper formed the basis for a large, multi-year joint industry project that provided a major advance in the understanding of pile behavior in clay at a fundamental level. The paper advanced the use of two new and highly innovative ideas in pile analysis, the cavity expansion process and Cam Clay theory, and it spawned a tremendous amount of industry research which ultimately led to the development of the improved pile capacity methods now in use.

Paper 3229 (1978) E.G. Ward, L.E. Borgman, V.J. Cardone

Statistics of Hurricane Waves in the Gulf of Mexico

This paper was in the vanguard of a new way to determine the likely future wave conditions by looking closely at the past, namely hindcasting. For the first time, using state of the art methods, a significant database of past hurricane wave statistics was developed, from which future return intervals could be inferred. The hindcast concept has been used in many geographical areas in the met ocean field, and has proven to be an excellent way to inform of expected future conditions. The information in this paper was contributory to various codes and standards. The hindcast concept has been used in many geographical areas in the met ocean field, and has proven to be an excellent way to inform of expected future conditions.

Paper 3380 (1979) P. H. Wirsching

Fatigue Reliability in Welded Joints of Offshore Structures

This paper created a format for assessing the fatigue reliability of offshore welded joints, focusing on tubular joints. The paper provides a means to evaluate risk in fatigue sensitive environments using little more than design data and the S-N fatigue design approach. This work has been broadly cited by other researchers and codes of practice, wherever fatigue reliability is discussed. The methodology has not changed to this day.

Paper 3690 (1980) J. A. Yura, N. Zettlemoyer and I. F. Edwards

Ultimate Capacity Equations for Tubular Joints

This paper was the most important advance in strength assessment/design of tubular joints since the early '70s, when all codified, offshore design practices were in their infancy. The paper's design equations greatly improved the reliability of joint designs and, at the same time, significantly reduced over-conservatism. The work spurred many worldwide follow-on studies. The recommendations were adopted by API and later were reflected in UK, Norwegian, and other practices. The equations were accepted, if not recommended, by all major Class societies and included in virtually all commonly used software packages. Several books refer to this work.

Paper 3951 (1981) L. J. A. Pinkster

Mean and Low Frequency Wave Forces on Semi Submersibles

This paper provided a new simulation technique which is nowadays widely used to calculate the mooring loads for semi submersibles in waves. It was the first time that diffraction theory was applied to calculate the drift loads on semi submersibles. The methods described in this paper are still widely used in the offshore industry to assess the mooring loads a semi submersible will encounter in waves.

2009 Award Papers — Volume Four

Paper 1899 (1973) R. G. Bea and P. Arnold

Movements and Forces Developed by Wave Induced Slides in Soft Clays

This paper describes the design methodology for permanent offshore structures to resist submarine slides. The methodology allows engineers to compute the soil movements and the forces imposed on the foundation piles during hurricanes caused by unstable sediments. The paper synthesizes field and laboratory data obtained at failed platform sites in South Pass. It describes the method for analyzing the force developed by the moving soil to illustrate the magnitude and depth of lateral soil movements. The method is still used to design platforms in deltaic areas.

Paper 1915 (1973) Gunnar Wold and Thorbjørn Kristoffersen

Development of Method for Measuring Susceptibility of Steel Plate to Lamellar Tearing

The paper presents a solution to an important issue in steel material properties that is critical to the design of deepwater structures. Once a lamellar tear occurs during construction it is very difficult to repair since most methods result in additional tearing. The paper describes a test method for assessing the susceptibility of steel plate to lamellar tearing. The method is still in use today and provided a solution that eliminated time consuming and costly repairs during construction.

Paper 2067 (1974) C. J. Garrison

Dynamic Response of Floating Bodies

The paper developed the first complete formulation and methodology to analyze the motions of floating structures with arbitrary geometry that are submerged or semi-submerged in water of finite depth. Computer methods are presented that determine the excitation forces and moments as well as added mass and damping coefficients for floating bodies. The method is still used today to analyze the response of large ocean structures floating in water of finite depth.

Paper 2671 (1976) I. M. Idriss, R. Dobry, E. H. Doyle and R.D. Singh

Behavior of Soft Clay Under Earthquake Loading Conditions

This paper presents the first described method for defining soil response degradation under earthquake loading in clay soils. The method is unique in that it allows modification of soil stiffness with time due to increase in pore pressure or remolding of clay soil. This innovative method provided the first opportunity to consider the change in soil response due to free field ground motions and the associated displacements.

The method has stood the test of time since it is incorporated in the DCHARM program that is still used to analyze offshore platforms located in seismically active areas.

Paper 3227 (1978) G. Z. Forristal, E. G. Ward, L. E. Borgman and V. J. Cardone
Storm Wave Kinematics
This paper was a groundbreaking effort that showed the importance of randomness and directionality in computing water velocities developed during hurricane waves. The paper was instrumental in establishing more accurate methods for determining wave and current forces on offshore structures. The method led to a reduction in wave drag forces on offshore platforms. The paper was used in the development of design standards and computer programs that are still used today in design practice.

Paper 3307 (1978) T. W. Miller, J. D. Murff and L. M. Kraft
Critical State Soil Mechanics Model of Soil Consolidation Stresses Around a Driven Pile
This paper confirmed that soil adjacent to a driven pile after installation will not reconsolidate along the critical state line. Hence, excess pore pressures need to be accounted for in models for axial pile capacity computations. The findings presented in this paper confirm that total stress methods can be used for pile capacity. The paper established the cornerstone of today's practice as used in the current methods to compute axial pile capacity. These methods appear in the API recommended practice and ISO standards confirming the paper's findings have endured the test of time.

Paper 4323 (1982) A. M. Reece and V. J. Cardone
Test of Wave Hindcast Model Results Against Measurements During Four Different Meteorological Systems
The paper established the validity of a hindcast method for developing storm wave design criteria. Knowledge of extreme wave heights is essential for the design of offshore structures. This hindcast model allows a wave database in one offshore region to be used to develop wave design criteria for other types of storms in other offshore areas of the world. The method serves as the basis for developing wave design criteria for most offshore areas worldwide.

2010 Award Papers — Volume Five

Paper 1263 (1970) J. R. Pauling and E. E. Horton
Analysis of the Tension Leg Stable Platform
This paper was historically important as the first exposure of the concept of the Tension Leg Platform (TLP)– a low- budget pioneering research effort –nearly a decade before the concept was widely accepted. The TLP is a type of moored stable platform whose buoyancy exceeds the weight. The supplementary equilibrating downward force is supplied by tensioned vertical tension cables or tendons. This paper gave proper emphases to avoiding slack in the tendons. The TLP suppresses the heave, roll and pitch motions. Design of a reliable anchoring system requires an accurate prediction of the forces in the tendons induced by wave action and the platform motions. The paper present a method for predicting the platform motions and forces in regular or random waves using a linear technique of hydrodynamic synthesis. Model experiments were conducted whose results display good agreement with predictions using this method. There are now three API RPs devoted to TLPs, and slender body modeling is still used for extreme wave preliminary design of TLPs.

Paper 2548 (1976) J. E. W. Wichers

On the Slow Motions of Tankers Moored to Single Point Mooring Systems

This paper was a groundbreaking effort that provided a complete technical solution to a complex mooring problem. It was one of the first to identify the underlying theory causing fishtailing of tanker moored to an offloading buoy. The paper summarized in a comprehensive manner the problem and the solution and articulated a new concept of stable and unstable equilibrium. The paper helped the industry improve design of a single point mooring (SPM). The technique described in this paper allowed designers, for the first time, to calculate the slow motions (fishtailing and galloping) of a tanker connected to a SPM. This new methodology changed the length and elasticity of bow hawsers during offloading operations to minimize incidents of tanker mooring failure for SPM systems. The importance of this work was realized immediately, and the paper was accepted for SPE journal publication. The principles of this research work are implemented in several major industry software tools. The work is the primary reference for all design codes that deal with SPMs. The low frequency viscous damping and the wave drift damping described are still widely used in time domain simulations for assessing stability of tanker connected to SPM systems.

Paper 2606 (1976) K. G. Nolte and J. E. Hansford

Closed-form Expressions for Determining the Fatigue Damage of Structures Due to Ocean Waves

This pioneering paper set the norm and established procedures for simplified fatigue analysis of offshore platforms. The paper made a milestone contribution in the evaluation of fatigue by providing a much needed approach to evaluate platform fatigue life, using a simplified approach that applied to the vast majority of platforms in the Gulf of Mexico. The method is used to evaluate the fatigue life to ensure adequacy of all major joints of a platform, and it is widely used by the offshore industry. It was adopted by API RP2A and is still the method for the conducting simplified fatigue analysis. As such it provided a significant advancement to safer design. The method has been incorporated into the leading industry software analysis, and today it still serves as the basis for a widely used, direct, relatively simple approach to control fatigue damage.

Paper 4115 (1981) V. Vivatrat and J. R. Kreider

Ice Force Prediction Using a Limited Driving Force Approach

This paper presents the first described method for limiting the driving forces of the ice pack pushing the ice feature against an arctic structure. Design loads can be limited by the driving forces of the ice pack pushing the ice feature into the structure, rather than the failure of the feature against the structure. Numerous flaws such as cracks, ridges, refrozen leads, and polynyas can reduce the force transfer capacity and induce non-uniformity in the stress field. The paper illustrates how a conical structure can create an ice rubble pile and describes methods of computing the driving force for structure design. The paper analyzes the different physical mechanisms that can lead to various loading scenarios in sea ice, leading to the first systematic examination if ice dynamics and loading processes. The paper had a major impact on future Arctic researches and designers considering the ice/structure interactions.

Paper 4429 (1982) J. A. Mercier, S. J. Leverette, and A. L. Bliault

Evaluation of Hutton TLP Response to Environmental Loads

This paper was a groundbreaking effort that described a holistic design process for the fist Tension Leg Platform (TLP)-Hutton. The process included new developments to meet the unique challenges of TLP response. The paper identified the worst cases of extreme met-ocean loads by combining analytical approaches with experiments to estimate the maximum values of unexpected response. The impact of various responses on design decisions in terms of performance, constructability, and operability were reported for the overall design of this first-of-a-kind structure. The findings identified important new design responses and learnings that are important and specific to TLPs, and critical for future designs. This major step forward tied theoretical loading/response concepts to practical realization of the hardware.

Paper 4465 (1983) A. G Young, G. W. Quiros, and C. J. Ehlers

Effects of Offshore Sampling and Testing on Undrained Soil Shear Strength

The paper was one of first and most comprehensive, to deal with the issue of offshore sampling and in situ/laboratory testing methods. Their influence upon sample quality, strength measurements, and interpretation of design strength profiles were clearly identified for improving the practice of foundation design. New push sampling and in situ testing methods were recommended that are now the standard practice recommended by API RP2A guidelines and ISO standards. The methods described in this work allow " high quality" deepwater boring data to be acquired and accurately interpreted for use with the current API empirical design methods. These methods have endured the test of time and are still referenced in the API recommended practice and ISO standard, confirming the paper's findings improved the practice of offshore foundation design.

Paper 4883 (1985) M. F. Randolph and B. S. Murphy

Shaft Capacity of Driven Piles in Clay

This paper was the culmination of over a decade of offshore industry sponsored work to produce new pile capacity methods for offshore use. Specifically this paper introduced a new method for calculation of shaft capacity of driven piles in clay. The method proposed in the paper contributed major advancements to the design of driven piles in clays. It differed from previous methods in that it allowed the effects of soil plasticity and stress history too be included via an analytical expression for the unit skin friction along the pile. It greatly improved upon the reliability of previous methods, was quickly recognized as the preferred method for the design of piles, and within two years was implemented in API recommended practices. It is still the method in the mail text of API RP2A and has been used extensively around the world, offshore and onshore, for the last 25 years.

Paper 4964 (1985) J. S. Templeton, J. D. Murff, R. H. Goodwin, and L. W. Klejbuk

Evaluating Soils and Hazards in the Mississippi Canyon

This paper describes a historically comprehensive inter-disciplinary study undertaken to clear obstacles to development in the complex geological environment of the deep water Gulf of Mexico. The effort marked the industry's first attempt to deal directly with the engineering problems posed by the distinctly different geologic conditions encountered beyond the continental shelf. The study combined geotechnical, geophysical and geologic studies with theoretical and numerical analysis as well as long term field and laboratory data acquisition programs. The study produced a revolutionary geologic interpretation for the Mississippi Canyon and the completely remove the obstacles to development. This study was the first to take advantage of the full benefits of integrated geophysical and geotechnical site investigations to resolve problems specific to complex deep water geology that might otherwise be prohibitive to development. The use of such integrated geophysical/geotechnical investigations has become the *de facto* standard for deepwater site evaluation. The 2,243-ft boring reported here is still a depth of penetration record for an offshore geotechnical site investigation.

2011 Award Papers — Volume Six

Paper 1898 (1973) G. H. Sterling & E. E. Strohbeck

The Failure of the South Pass 70 "B" Platform in Hurricane Camille

This paper describes the failure of a newly installed platform in about 100m of water during Hurricane Camille in 1969. The failure mechanism was hurricane wave pressures on the seabed that caused the softy clay soils to move and topple the structure. This was the first publication which presented field data on the effect of hurricane wave induced pressures on soils and resulted in significant research on the subject and changes to design practice in similar environments. Little attention had previously been paid to geological hazards for deepwater structures as a design constraint. This paper was one of the first to set a standard for describing engineering failures in the offshore environment by completely documenting the phenomenon so that others might learn from the failure. The openness of the paper in documenting the failure is admired by the profession and serves as an example to the industry that failures can be an important learning tool if shared by all.

Paper 3377 (1979) R. B. Reimer, R. W. Litton and J. H. Babcock

Finite Element Analysis of Complex, Welded Tubular Joints

This paper outlines an exemplary Joint industry Project to automate the use of finite element analysis to model complex tubular joints, in which actual 3-D weld dimensions are used. The model uses isoparametric curved shell elements that allow a linear trend of stresses and strains towards the weld toe. The model provides hot spot stress data consistent with empirical hot spot S-N curves presented in API, ISO, IIW design standards. The automated results supported parametric studies that lead to formulas for stress concentration factors that are included in these standards. The JIP made an outstanding contribution to the practice of offshore civil engineering by providing new data, new analytical methods, new structural system hardware, new insights into existing problems, and a durable approach to applying fatigue data and analysis methods.

Paper 3492 (1979) R. F. Scott

Cyclic Static Model Pile Tests in a Centrifuge

This paper showed the technical feasibility and attractiveness of using centrifuge model testing to solve complex soil-structure interaction problems such as pile-structure response to cyclic axial and lateral loading. The paper confirmed that centrifuge model testing overcomes many of the problems associated with field-testing or small-scale model testing by testing a small model at prototype level stresses and strains. The soil response can be correctly modeled in the centrifuge by performing the test at a higher gravitational acceleration than a 1 g model. The paper confirmed the important role of centrifuge testing in offshore geotechnical engineering for solving complex problems that would be costly and difficult by other ways. The paper added and important new tool to the geotechnical engineer's tool kit for all types of offshore foundations. The paper established the centrifuge as a legitimate tool for analyzing full-scale foundations under complex loadings. The paper provided an impetus for research groups around the world to invest in the construction of centrifuges that have been essential in improving the practice of offshore foundation design.

Paper 3764 (1980) Ahsan Kareem

Dynamic Effects of Wind on Offshore Structures

This paper provided a thorough, technical basis for the methods adopted to calculate dynamic wind loads on compliant structures. Their long period response to environmental forcing was under scrutiny. This paper laid out the procedure and the principles needed to address this significant design element. This paper describes the stochastic nature of a wind field and demonstrates how its coupling to structure responses could be quantified. It describes the relationship between the second order features in the wind to long period responses of a compliant structure. It demonstrated the existence of forcing that contributes relevantly to structure design for resistance to extreme loading events. This paper provided a though, technical basis for the methods adopted to calculate dynamic wind loads on compliant structures. The paper is referenced in the current version of API RP 2T.

Paper 3821 (1980) Stephan A. Mahin, Egor P. Popov, and Victor A. Zayas
Seismic Bahavior of Tubular Steel Offshore Platforms

This paper presented a fundamental advancement in the structural design of offshore platforms subject to severe earthquakes. The results of large-scale tests demonstrated the importance of platform ductility and energy absorption, critical for adequate performance under severe earthquake motions. Test results confirmed the importance of proper d/t ratios of critical platform structural elements. It also confirmed the importance of, as well as the ability of, platform structures to perform in a ductile manner, allowing energy absorption capacity to be preserved well into the inelastic range. The paper provides insight into performance of typical X braced platform framing, with special attention to the importance of d/t in controlling local as well as column buckling. It also noted that the hysteretic behavior of buckled braces resembles that for individual braces with fixed ends rather that pinned. The value of the paper is significant in providing quantitative as well as qualitative insight in evaluating the importance of ductility in the performance of platform structures, specifically under severe earthquake loading.

Paper 4178 (1982) Vitoon Vivatrat, Phiplip J. Valent, and Albert A. Pointerio
The Influence of Chain Friction on Anchor Pile Design

This paper is a fundamental contribution to mooring line analysis. The paper presents for the first time the non-linear governing differential equations for soil-mooring line interaction, provides recommendation for characterizing the soil resistance and anchor line, and describes a practical method for solving the equations. The paper has had the broadest applications. It was originally developed for anchor pile analysis but it is now used in the design of virtually all anchor systems that involve mooring lines. Among these applications are suction caissons, drag anchors, VLAs, and SEPLAs. The paper is a basic reference in the literature and in codes and practices.

Paper 5096 (1986) G. Rodenbush
Forces on a Large Cylinder in Random Two-Dimensional Flows

This paper describes a novel laboratory investigation in which a large-scale cylinder was moved through the water in a controlled manner to realistically simulate wave-induced flow past cylindrical members on a platform under design conditions. In addition to providing large scale (Reynolds number grater than 10^6, the experiments investigated the effects of two-dimensional flow, randomness, currents in addition to wave-induced flows, and cylinder roughness. It provided a unique data set for computing the drag and inertial coefficients required to compute design wave forces on a platform. The paper also compared these results with various existing field measurements of forces on platforms and other experimental data. The overall results provided a unifying description of the force coefficients for computing wave forces on platforms. These data and the comparison and correlation with the full-scale field data provided a strong basis for the wave force coefficients that are used to design offshore platforms, and improved the reliability and confidence in design wave forces. These results are still valid and in use today.

2012 Award Papers — Volume Seven

Paper 3817 (1980) J. R. Hogervorst

Field Trials with Large Diameter Suction Piles

This paper showed for the first time that large diameter piles (i.e., caissons) could be installed by penetrating into the seabed by suction alone. The author presents design equations for the installation forces and the holding capacity of the pile to lateral pulling forces and compares them to field trial results. This paper demonstrated its practicality for use by the oil industry. Suction piles (i.e., suction caissons) are now widely used to temporarily anchor exploratory drilling rigs and as permanent moorings for deepwater production platforms such as SPARS and floating production systems. This paper confirmed the suction pile concept, and subsequently the technology was licensed to SBM for use as a temporary mooring system for FPSO ships and exploratory drill rigs. Subsequently, the concept was applied to permanent mooring systems for deepwater production platforms. It is now the most common foundation method for permanent moorings in deepwater except for TLPs where driven tension piles are still used.

Paper 4119 (1981) David. B. Prior, James M. Coleman, and Louis E. Garrison

Geologic Mapping for Offshore Engineering, Mississippi

This study was the first to comprehensively describe near-surface processes off the Mississippi Delta. For the first time engineers could understand where and why soil movements occur off the Delta. This information allowed structures to be safely designed to resist the soil movements that occur off the Delta. A uniquely detailed geophysical survey was completed in 1977-79 for 100 lease blocks (900 square miles) in the mudslide area off the Mississippi Delta in water depths of 30-800 ft. The resulting maps allowed an assessment of the location, magnitude, and frequency of seafloor processes for siting and design of structures and pipelines. The near-surface geology and instability distribution maps were prepared from high-resolution geophysical and side-scan sonar surveys and supplemented by precision bathymetry, bottom samples, and data from soil borings. The methods of survey and data synthesis are applicable to other areas where similar engineering constraints may exist. The methodology used in this paper is the basic framework for all integrated geoscience studies conducted for structure placement. All structures placed within that geological province use the results given in this paper.

Paper 5103 (1986) T. I. Tjelta, T. R. Guttormsen, and J. Hermstad

Large Scale Penetration Test at a Deepwater Site

This paper documents the feasibility of deeply penetrating (22m) a thick (0.4m) concrete caisson skirt into an interbedded clay, clayey sand and sand profile using suction as a driving force. This critical field test provided key elements for the design of the massive Gulfaks and later Troll structures, two of the largest structures ever built. Furthermore the data derived from the test have been widely used for suction caisson design for a wide range of applications including deepwater anchors. This major field test, at near prototype scale, was a key element in proving the viability of using gravity structures in a "soft soil" environment. The foundation instrumentation included 13 types of instruments with a total of 70 sensors and provided major new insights into the mechanics of suction caisson installation. The paper has been widely used in practice for establishing parameters needed to predict suction caisson penetration resistance particularly for clay profiles with interbedded sands. The test has been cited in recent years some 20 plus years after publication. More importantly the results of the test are widely known in the profession and the general knowledge of the test outcomes has provided geotechnical engineers with the confidence that only a test at such a large scale can bring.

Paper 5407 (1987) John H. Pelletier and George E. Sgouros

Shear Transfer Behavior of a 30-in. Pile in Silty Clay

This paper provided a major addition to understanding pile to soil shear (T-Z) transfer behavior, being one of only a handful of pile capacity tests at near prototype scale and the only large scale test in an important subset of soil conditions, over consolidated silty clay. During the pile test T-Z curves were measured along the length of the pile. Using beam-column theory the measured pile-top load deflection behavior under both static and cyclic loading can accurately be predicted with the measured T-Z curves. In addition to the monotonic T-Z results, the insight provided by the paper regarding the difficult issues of cyclic loading and rate effects is of major importance. The paper is widely cited in the extensive technical literature on load test data analysis as well as in most offshore foundation design practices. The Beta pile load test remains a pillar of the offshore foundation design practice. Its influence continues to be felt and will continue to be an important asset indefinitely.

Paper 5524 (1987) J. K. Vandiver and T. Y. Chung

Hydrodynamic Damping on Flexible Cylinders in Sheared Flow

The paper provided the first experimental quantification of the influence of sheared flow on vortex induced vibration of slender structures such as risers. The quantification of hydrodynamic modal damping provided experimental justification for the hypothesized yet unproven high damping of vortex induced vibrations (VIV) on slender structures when exposed to realistic non-uniform, i.e. sheared, current profiles. This work reduced the conservatism required for the VIV design of steel risers in deepwater thus bolstering the development of this critical technology which is broadly used and features in most deepwater developments today. This work helped launch additional experiments to further probe the influence of shear on the VIV response of slender bodies in realistic ocean currents. This paper describes key elements of the theoretical basis and experimental support for a computer program and design tool, SHEAR7, which is widely used in the industry to characterize the effects of VIV on long slender structures such as risers. This is arguably the industry standard for this sort of calculation. This paper helped advance the design of deepwater risers that are ubiquitous in the offshore industry. The concepts and observations contained therein are still relevant today and are incorporated in design tools still being used in the industry.

Paper 5555 (1987) W. H. Luyties and J. F. Geyer

The Development of Allowable Fatigue Stresses in API RP2A

This paper provides the basis and explanation of a simplified fatigue procedure that has endured through many subsequent editions of RP 2A. The study served to calibrate the new simplified fatigue design procedure. Detailed fatigue analyses were performed on seven Gulf of Mexico structures with different geometries, natural periods, and water depths ranging from 58 to 314 feet. One of the powerful features of the procedure is that wave climate is explicitly included in its formulation. Thus, when the stress range at design loading is calibrated to its counterpart in the generic fatigue analysis, it is possible to develop a preliminary allowable stress for other cyclic loading environments, using the Weibull statistical parameters derived from the generic fatigue analysis in that same environment. Beyond fixed platforms in the Gulf of Mexico, the general method has been extended to apply to compliant towers, TLPs, tubular highway bridges, and the movable roof in a domed stadium.

2013 Award Papers — Volume Eight

Paper 3027 (1977) Svein Fjeld

Reliability of Offshore Structures

This was the first OTC paper to treat the reliability of offshore structures in a general way, covering loads and resistance for structures and foundations of all types. It was originally, and still is, a cited reference in API RP2A LRFD. It has been cited in subsequent OTC papers as the significant reference on the subject. This was the landmark paper that described "overarching role" that reliability analysis plays in the design of offshore structures. It set the stage for a large body of future work in this area.

Paper 3046 (1978) F. Moses and B Stahl

Reliability Analysis Format for Offshore Structures

This was a "masterpiece" paper that established the method for using progressive collapse (unzipping) analysis to determine not only the expected value but also the variance of the structural resistance. The paper provided the basis for performing calibration computations needed to set load and resistance factors so the LRFD-based design that would result in platform reliability levels consistent with prior designs. Of all the OTC papers related to reliability, this paper was the most innovative in providing a method to extract variance as well as mean values for structures from a straightforward structural analysis. The method in turn opened the door to a decade of reliability calibration work that would eventually underpin API RP2A LRFD. The paper was originally, and still is, a cited reference in API RP2A LRFD.

Paper 5699 (1988) F. Moses and R.D. Larrabee

Calibration of the Draft RP2A-LRFD for Fixed Platforms

This paper addressed the reliability calibration important to determining the principal safety factors that formed the basis of the API RP2A LRFD design for offshore platforms. This is fundamental to the successful use of LRFD as a design tool. The results and recommendations of this paper form the basis for LRFD design that is common practice wherever this method is used. The paper contributed significantly to the safe design of platforms by providing a rational basis to support the important factors essential for safe and economical design of platforms, based on prior experience with WSD practice. The paper was originally, and still is, a cited reference in API RP2A LRFD.

Paper 5882 (1988) J.R. Lloyd and D. I Karsan

Development of a Reliability-Based Alternative to API RP2A

This paper provides classical references, historical basis, and motivation to develop the LRFD design method. The paper is a companion to OTC 5699 by providing summary results and a guide to practical implementation of numerical calibration studies. It describes the main differences between LRFD (load & resistance factor design) and WSD (working stress design). The paper also discusses the philosophical reasons for making the change and laid out the strategy for implementation. After review and revision, API RP2A-LRFD as described in this paper was adopted in its entirety for the first ISO international standard for fixed steel offshore platforms. It has been used as such, worldwide, for twenty years. The method as described in this paper has served as the template for development of ISO 19902, recently issued after fifteen years of further development. It is one of the options implemented in most design software, such as SACS, MicroSAS, etc. The API RP2A-LRFD standard envisioned in OTC 5882 was the culmination of a ten-year JIP effort in API, and the impetus for a subsequent fifteen-year effort in ISO. It was prescient in identifying the vulnerability of deep shelf fixed platforms designed by WSD, painfully revealed by hurricanes Ivan, Katrina, Rita, and Ike, and now remediated by new reserve strength checks. This paper was originally, and still is, a cited reference in API RP2A LRFD.

Paper 6001 (1989) S. Lacasse and A.M. Goulois

Uncertainty in API Parameters for Prediction of Axial Capacity of Driven Piles in Sand

The paper made a key contribution to understanding the bias and uncertainties in axial pile design in sands needed for development of portions of RP2A LRFD. The papers compiled the judgment of practicing geotechnical engineers relative to the probability density functions of six key variables needed to compute the axial pile capacity of piles. The paper presented the statistical compilation of 31 expert opinions relative to the major biases and uncertainties to be compared with pile load test results. The average expert opinions allowed geotechnical engineers to understand if the API RP2A method is a best estimate or a conservative threshold. The paper greatly improved the reliability of offshore pile design and influenced the selection of soil parameters used in the API recommended practice.

Paper 6379 (1990) W. H. Tang, D. L. Woodford, and J. H. Pelletier

Performance Reliability of Offshore Piles

This paper was a major contribution to the development of the foundation section of API RP2A LRFD by providing a comprehensive statistical analysis of pile load test results in clay and sand for both axial and lateral loading. It also identified and quantified the many effects that influence pile response and testing results such as consolidation, cyclic loading, rate effects, and random variability of sampling test data due to disturbance. The paper provided a systematic study of all aspects of pile design needed to formulate pile design practice for RP2A LRFD. In the long term the study facilitated the development of pile design procedures that maintain a consistent level of safety for different site conditions, alternative pile configurations, and structural and foundation subsystems.

2014 Award Papers — Volume Nine

Paper 6174 (1989) B.L Grundmeier, R.B Campbell and B.D Wesselink

A Solution for Wind Induced Vortex Shedding Vibration of the Harmony and Heritage Platforms during Transpacific Tow

Wind induced vortex shedding vibration of steel tubulars in offshore jackets was a growing problem in the late 80's. In several cases fatigue cracking at joints had occurred both during fabrication and transportation, particularly in long or stormy tows. The prevailing methods to evaluate susceptibility to vibration and damage were inadequate and sometimes misleading. VIV in air is quite different from that in water, due to the thousand-fold difference in medium density versus object mass. Based on industry anecdotal experiences, coupled with hard data from Shell's Bullwinkle platform, Exxon developed methods to avoid the issue on their own transpacific tow, providing an improved basis for dealing with a very difficult problem. This is a practical application of 'execution' oriented engineering to help project engineers deal with events that are hard to understand and predict, but very damaging to assets and careers if they occur. Much of this paper's approach found its way into industry practice.

Paper 6376 (1990) J. D. Bogard and H. Matlock

Application of Model Pile Tests to Axial Pile Design

This paper presented a practical empirical method to calculate the time-dependent setup of driven piles. The first practical application of the methodology was in the 1996 installation of the Mars Tension Leg

Platform (TLP) where the driven piles were directly connected to the tendons and where the tendon installation directly followed the installation of the driven piles. Normal practice prior to this installation was to allow enough time for the piles to setup which required two field installations and potential delays in starting production. The effect of this strategy was a significant cost savings. The direct use of setup relationships is now a standard design practice for driven piles used in TLPs.

Paper 6387 (1990) P. W. Marshall, J.W. Post, R.E. Frischmuth, and J.L. Grover
Advanced Fracture Control Procedures for Deepwater Platforms and Compliant Towers
This paper set the standard procedure used to control fracture for thick walled steel joints of offshore structures in deeper waters. The procedure was based on real-world tests performed on the Cognac and Bullwinkle platforms, respectively, installed in 1020 and 1353 feet of water depth in the Gulf of Mexico. The methods and procedures developed for this effort were then used in developing the U.S national codes for tubular structures. These include AWS D1.1, API RP2A, and API Specs 2H, 2W, 2X, 2Y -- as well as related international standards for steel, welding and testing, as evidenced in the Material Category approach in ISO 19902. In addition to fixed platforms, these concepts and methodologies have been successfully implemented in the design and construction of TLPs, Compliant Towers, Production Semisubmersibles and Topsides. They have also been implemented in tubular sections of highway bridges. The paper led to follow-up work at the University of Newcastle-Upon-Tyne, disseminating strategies for dealing with structural integrity and fracture control of thick walled steel joints, high cost materials, and addressing a plethora of concerns from the highly technical to the daily QA/QC controls from weld flaws, to personnel qualifications for assessing quality.

Paper 7196 (1993) W. H. Tang and R.B. Gilbert
Case Study of Offshore Pile System Reliability
This paper made a key contribution to the understanding of pile system performance. It provided important insights as to the effects of the pile foundation geometry including pile layout, number of piles, and ratio of water depth to base width as well as the role of dominant failure mode on the reliability of a system designed per current API RP2A. The paper documented in the public domain the findings of the authors' API sponsored study and made a significant contribution to formulating the requirements for platform requalification. The work has been used extensively in more recent studies attempting to calibrate foundation performance in light of damage assessments subsequent to severe hurricane exposure.

Paper 7495 (1994) M. L. Peterson and E. L. von Rosenberg
Offshore Platform Steel Specification
The technology summarized in this paper resolved some worrisome technical issues, and introduced a step change in the steel selection and fracture control practices for fixed and compliant offshore structures in deepwater. The paper helped establish a program for fracture mechanics testing as specified in RP2Z to verify that any reduction of fracture toughness resulting from welding is within acceptable limits with welding procedures used by the fabricator. Verification of the acceptability of the steel before it is produced and delivered will avoid the risk of receiving material that cannot be economically welded under typical fabrication yard conditions. The paper also provided a summary of the technology for using high performance steel in critical offshore structures subject to fatigue, as it has been applied for the last 2+ decades, and adopted into international standards such as ISO 19902. The less onerous preheat requirements of advanced MP steel (API Spec 2W and 2Y) versus conventional CP steel, under high and moderate restraint welding conditions, explains their popularity for high-strength heavy-thickness applications.

Paper 7620 (1994) *E.H. Phifer, Frans Kopp, R.C. Swanson, D.W. Allen, and C.G. Langner*

Design and Installation of Auger Steel Catenary Risers

This paper provides a complete review of problems and solutions in the design and installation of the first steel pipe catenary risers (SCRs). It addresses structural modeling, weather modeling, material and welding requirements, flexible joint connection, and several practical installation, transfer and pull in topics, with a clear exposition of the issues and the solutions actually used. Since the installation of those SCRs in 1994, there have been numerous further developments and applications including refinements of design analysis tools; extension of SCR designs for applications to Spars, Semi-FPUs, and ship-shaped FPSO platforms; extension of SCR installation techniques to include S-Lay, I-Tube, and 1st End installations; further development of stress joints, I-tube hang-off techniques, and VIV suppression. Over 200 major risers have been installed worldwide as of 2014 as a result of the technical contributions made by this paper. In addition, it has facilitated the use of many smaller diameter SCRs as import risers to tie-in remote subsea wells.

2015 Award Papers — Volume 10

Paper 5346 (1986) J.G. Mayfield and M.E. Zimmerman

Notes on Heavy Lift

The paper describes extensively the interrelated activities and responsibilities of the various organizations involved in the design of lifting systems and components for the offshore installation of topside facilities of oil and gas platforms. The subject matter is treated in detail—module weight, padeye designs for lifting ease, cable laid slings, positioning of slings, guide and bumper systems, rub strips—a virtual godsend for the installation team. The most significant contribution of the paper involves the selective positioning of slings, based on measured lengths, to obtain the most favorable load distribution for heavy lifts. In addition the paper provides commentary based on years of experience and practical applications to the representatives who must coordinate, monitor and approve the activities of the organizations involved. The result is a practical guide helpful in planning for the execution of all types of project lifts. The paper helped assure that offshore heavy lifts worldwide were safe, successful and performed in a timely fashion.

Paper 6615 (1991) J.J. Osborne, J.C. Trickey, G.T. Houlsby, and R.G. James

Findings from a Joint Industry Study on Foundation Fixity of Jackup Units

This paper is, historically, the most significant paper in the area of jack-up foundation assessment for storm loading conditions. The paper presents the results of a JIP as to how to account for rotational fixity of jackup rig foundations. All the foundation-related information referenced in SNAME Bulletin 5-5 and ISO 19905 were built on the groundwork laid down in this JIP. The paper was the first publication to present a comprehensive and technically defensible method for assessing the foundation performance under storm loading conditions. The paper describes an approach for developing a foundation yield interaction surface that is now referenced in most foundation standards and practices.

Paper 6757 (1991) J.R. Fowler and C.G. Langer

Performance Limits for Deepwater Pipelines

This paper highlights the differences between onshore and offshore pipeline design, specifically for deep water applications and presents experimentally validated formulae for buckle propagation, collapse due to external pressure with and without bending, as well as allowable stresses. The paper was the first that described the special design challenges of deepwater pipeline installation. The formulations in this paper were written in 1991, and codified in API RP 1111, well before the first deepwater pipelines were installed.

Government regulators all across the globe accept the use of API RP 1111. The equations in this paper have stood the test of time, and more complex equations in other guidelines remain firmly based on the same equations presented in this 1991 paper. As the subsea industry moves ever more into deeper and higher pressure developments, the guidelines presented in this paper continue to be as relevant today and for the foreseeable future.

Paper 7153 (1993) C. Petraukas, J.C. Heideman, and E.P. Berek
Extreme Wave-Force Calculation Procedures for 20th Edition of API-RP-2A
This paper presented a significant revision of the guideline design wave force calculation procedure recommended in the 20th Edition of API RP2A WSD and the 1st Edition of API RP2A LRFD. The revision incorporated the results of research and experience obtained over the previous 25 years, in particular advances related to force coefficients and wave and current kinematics. The paper also described the calibration of this procedure against historical platform experience. The paper provided engineers involved in the design and assessment of offshore structures with an important introduction to the significant changes in the new procedure and the bases for these changes.

Paper 7306 (1993) D.E. Jones, W.T. Bennett, and M.J.R. Hoyle
The Joint Industry Development of a Recommended Practice for the Site Specific Assessment of Mobile Jackup Units
This paper served as a major advancement in the establishment of an industry practice to evaluate site-specific parameters needed to assure jack-up foundations are suitable for a site. It helped establish consistent and generally applicable criteria for the assessment of jack-up unit foundations. The methodology incorporated all considerations including environmental loadings, structural analysis and assessment practice. The methodology described in this paper forms the basis of jack-up assessment today and is referenced in the ISO standard. The methodology unified many diverse practices for assessment of jack-up foundations and helped move the industry toward a consistent approach. Such a consistent approach is important to assure that jack-ups would be operated within their design limits in order to reduce the number of accidents and fatalities associated with jack-up foundation failure.

Paper 7668 (1995) R.N. Dutt, E.H. Doyle, J.T. Collins, and P. Ganguly
A Simple Model to Predict Soil Resistance to Driving for Long Piles in Deep Water Normally Consolidated Clays
The paper presents a model for predicting the drivability of large-diameter, long piles in normally consolidated clays as would be used for foundations to support TLP's and SPAR's. Case studies at the Bullwinkle, Auger, Pompano, and Ewing Banks platforms were used to calibrate the model and to illustrate that adequacy of the new model. Subsequent studies have verified the accuracy of the method, so the method provides more assurance in our capability to predict the drivability of large diameter pipe piles. The method is considered by ABS to be preferred procedure for predicting pile drivability and is referenced in the API–2SK. The method has been used for a majority of the deepwater TLP's in the Gulf of Mexico. The improvement in the prediction of pile drivability provided by the method also allows an improved prediction of pile fatigue which is a critical component in the design of long piles as used to support TLP's.

2016 Award Papers — Volume 11

Paper 6541 (1991) G.Z. Forristall, R.D. Larabee, and R.S. Mercier

Combined Oceanographic Criteria for Deepwater Structures in the Gulf of Mexico

This paper introduced the concept of response-based criteria focusing on the maximum structural responses rather than just the maximum metocean parameters in selecting the design criteria. The approach addresses both the development of design criteria for structures in complex metocean environments (wind, wave, and currents that are non-collinear with non-concurrent maxima) and structures with multiple elements that each can reach their maximum design response in different combinations of winds, waves, and currents. The approach also incorporated hindcast metocean conditions for a historical population of storms so as to account for the expected variability and uncertainties in the environment and the structure's responses during its lifetime. The paper has been cited in more than 40 publications that address developing design criteria for different fixed and floating structures in a variety of metocean conditions.

Paper 7157 (1993) J.H. Pelletier, J.D. Murff, and A.G Young

Historical Development and Assessment of the Current API Design Methods for Axially Loaded Pipes

The paper presents the most extensive account of the evolution of the API axially loaded pile design methods ever published. It made a significant contribution to the offshore industry by condensing and presenting the salient points of more than 60 studies. The paper should not be confused with a mere "state of the practice summary, since the authors not only describe the "what" of the methods but also give their views and insights as to the "why". The paper tells the story of how the offshore pile design methods evolved from an insider point of view. The authors also voluntarily point out the limitations of the API methods and give recommendations for future work and research. The paper has been widely used and is referenced in API RP 2GEO. The International Society of Soil Mechanics and Geotechnical Engineering (ISSMGE) and its Technical Committee 209 on Offshore Geotechnics also reference the paper. Their website lists it as one of the top 14 key conference papers of all time in offshore geotechnics.

Paper 7237 (1993) Erling Huse

Interaction in Deep-Sea Riser Arrays

This paper describes the first practical model for predicting drag forces on arrays of cylinders and a methodology for identifying the interaction (e.g. collision) between risers. Riser interference is a key design consideration for all deepwater floating production systems. The paper introduced the first practical model that could assess the potential for riser classing. The author combined classical hydrodynamics and engineering judgment to develop his proposed approach. The results were compared with published model test results for fixed arrays of cylinders and agreed remarkably well. The numerical procedure described in this paper is still in use today. It has been implemented in commercial riser analysis software and the "Huse model" is cited in codes and practices that include APR 2RD and DnV RP-F203 (Riser Interference).

Paper 7482 (1994) W.F. Krieger, Hugh Bannon, J.R. Lloyd, R.S. De, K.A. Digre, Damo Nair, J.T Irick, and S.J. Guynes

Process for Assessment of Existing Platforms to Determine Their Fitness for Purpose

The impetus for developing assessment guidelines was the evolution of platform design practice over the past forty years, resulting in new platform design standards more stringent than those used earlier. Concerns were then raised regarding the adequacy of older structures. An API task group was formed to develop guidance on assessment of existing fixed offshore platforms. The paper explains and justifies the assessment process. The criteria are based on over forty years of successful operations, field experience, and detailed investigation of platform failures and survivals in past hurricanes. Two levels of analysis may be used, with increasing complexity and decreasing conservatism design level and ultimate strength. Because of the wide sweep of its content, experienced industry experts from wide technical and experimental backgrounds contributed to the study and to the paper.

Paper 7484 (1994) Charles Petrauskas, T.D. Finnigan, J.C. Heideman, M. Vogel, M. Santala, and G.P. Berek

Metocean Criteria/Loads for Use in Assessment of Existing Offshore Platforms

This paper addressed the development of metocean criteria for the assessment of existing platforms based on one of four exposure categories and the type of storm system relevant to each. The development of "sudden storm" criteria allowed for a realistic approach to creating "fit for purpose" criteria. The results and procedures presented involved a consensus approach on how to arrive at the appropriate metocean parameters for each exposure category and provided a unique approach to the assessment of existing structures. The paper also included a simplified procedure for calculating wave-in-deck forces that is still in use. The work associated with this paper provided the basis for the draft API document "RP2A-WSD, Section 17, Assessment of Existing Platforms". The procedures outlined in the paper were widely used for many years in addressing "fit for purpose" issues and the paper is still referenced (RP2-SIM).

Paper 7988 (1996) H.H. Roberts, E.H. Doyle, J.R. Booth, B.J. Clark, M.J. Kaluza, and A. Hartsook

3D-Seismic Amplitude Analysis of the Sea Floor: An Important Interpretive Method for Improved Geohazards Evaluations

The use of lower resolution 3-D exploration-level data to replace high-resolution geophysical data was intuitively the wrong direction to go in understanding and identifying potential geological hazards. However, through the use of 3-D data to construct bathymetry charts and amplitude renderings, seafloor features became quite clear using this data set as demonstrated in the paper. This paper uses several data examples to show that the procedure is an effective tool to understand geological features. By demonstrating the use of 3-D exploration-level data to identify surface features, 3-D data was proven to be a valuable data set that could be adapted to the workstation environment for mapping subsurface features. In going from acquiring high-resolution data to using readily available 3-D data, the cost of geological hazard investigations was significantly reduced as was the time to produce a report. The use of 3-D data changed the way industry did geological hazard investigations. It became the standard for all geological hazard investigations worldwide.

2017 Award Papers — Volume 12

Paper 6127 (1989) J.N. Brekke, J.D. Murff, R.B. Campbell, and W.C. Lamb

Calibration of Jackup Leg Foundation Model Using Full-Scale Structural Measurements

The paper described how structural measurements made from an operating jack-up during a storm were used to calibrate jack-up foundation models. The work was very innovative since it provided an opportunity to calibrate soil structure models with dynamic analysis results based on field measurements during a hurricane in the Gulf of Mexico. The measurements show that, for significant leg penetration in soft clay, the foundation affords considerably more benefit to foundation fixity than previously understood—over 20 percent in the field case investigated. The work described is this paper has been widely used and referenced in later studies. It is still used for jack-up site-specific assessments today. The methods provided in this paper are cited in current issues of SNAME Bulletin 5-5 and ISO 19905.

Paper 6335 (1990) A.C. Palmer, C.P. Ellinas, D.M. Richards, and J. Guijt
Design of Submarine Pipelines Against Upheaval Buckling
Upheaval buckling can occur under certain conditions of high temperature and pressure where the pipe moves upward and out of its trench, thus exposing this portion of the pipeline to external hazards. The paper addresses the problem of designing the trench and overburden for a buried subsea pipeline to prevent upheaval buckling. The paper provided a major advancement to prevent upheaval buckling allowing the oil industry to work in the North Sea and Arctic regions along with the fishing industry. The work described is this paper has been widely used and is still relevant to pipeline design today. The methods provided in this paper are cited in current issues of API and DNV Recommended Practices.

Paper 6831 (1992) C.K. Cooper
A Preliminary Case for the Existence of Hurricane Alleys in the Gulf of Mexico
The paper describes the existence and rationale for hurricane alleys and the spatial variability in design wave heights along the Gulf Coast. The study showed a strong correlation between spatial variability in extreme wave heights along the coast with warmer water temperatures. The work was important in the evolution of Gulf of Mexico design criteria. The variation in design wave heights was incorporated in API Bulletin on Interim Guidance on Hurricane Conditions in the Gulf of Mexico 2INT-MET (2007) and in API RP 2 MET 1st Edition (2014).

Paper 6965 (1992) C.R. Chaplin and C.J.M. Del Vecchio
Appraisal of Lightweight Moorings for Deep Water
The paper discusses the advantages of using lightweight fiber ropes for mooring deep water floating production platforms. The JIP study described the modeling of polyester mooring systems that enabled the design and installation of the first polyester mooring system in the world installed in 1997. The study demonstrated the importance of selecting mooring line stiffness judiciously so as to meet design requirements for station keeping, strength, fatigue, creep, and wear while minimizing overall system cost. The paper—after 20 years—is still referenced in publications describing the design of polyester mooring systems.

Paper 8452 (1997) R. Nelson, J. Soliah, T. Caldwell, D. Morrison, and J. Pritchard
Heave Compensated Landing System – A Novel Tool for Subsea Intervention
The paper describes a new intervention system that allows the use of a low-cost workboat to deploy and recover delicate payloads during subsea installations. The innovative system decouples vessel motions from the subsea package by using a specific chain and buoy system. The concept for the Heave Compensated Landing System (HCLS) is presented along with the offshore testing program, test results, lessons learned, and potential future developments. The system is truly a significant innovation that combined intuitive field engineering with in-depth analysis. The methodology has become "the way" subsea payloads are routinely installed. The methodology has become the method used for offshore operations that substantially reduces installation costs.

Contents

OTC 6127

Calibration of Jackup Leg Foundation Model Using Full-Scale Structural Measurements

by J.N. Brekke, J.D. Murff, R.B. Campbell, and W.C. Lamb, Exxon Production Research Co.

This paper was presented at the 21st Annual OTC in Houston, Texas, May 1–4, 1989.

ABSTRACT

The fitness of jack-up rigs to operate in varying water depths and weather conditions is determined by "site-specific" structural assessments. These assessments are conducted by a wide range of companies which includes oil companies, rig owners, rig designers, classification societies, warranty companies, and regulators. The results of a particular assessment can differ significantly, depending on the assumptions and modeling details used in the structural analysis. Key aspects of the analysis include the leg foundation model, the calculation of wave load, and dynamic amplification effects.

This paper describes how structural measurements during a storm on an operating jack-up rig were used to calibrate a leg foundation model. The measurements show that, for significant leg penetration in soft clay, the foundation affords considerably more structural resistance, or "fixity", than is assumed in the typical, "pinned spudcan" model. When applied to an assessment, the revised leg foundation model leads to a 20 percent reduction in the maximum leg moment and an additional 20 percent reduction in the structure's natural period which further reduces the load.

INTRODUCTION

Jack-up rigs are self-elevating mobile drilling units which are used extensively offshore to drill both exploratory and production wells. Unlike most offshore structures, a jack-up typically operates at a given site for only a few months before moving. The unit is emplaced by floating it over a location and lowering its legs until the spud cans contact the seabed. Following proof-testing of the foundation (pre-loading), the hull is elevated to provide a stable platform from which drilling operations can be conducted.

In an effort to resolve some of the uncertainties in the analytical idealization of jack-ups, a field measurement program was undertaken. Measurements were made during February-May, 1988 on the KEYES 301 jack-up (rig type L-780 Mod II) while it was operating in the Gulf of Mexico. Structural response and metocean data were collected during a storm in which the largest waves measured 25 feet from crest to trough. The data were processed to determine 1) reaction loads at two elevations on one of the three legs, 2) motion of the hull, and 3) metocean conditions. The measured loads and motions were then compared with analytical predictions based on the measured metocean conditions.

This paper focuses on the calibration of a leg foundation model with the measured structural response data. The text includes 1) a summary of the measurements, with emphasis on leg foundation fixity, 2) a description of the leg foundation model and its basis, 3) comparison of measured and predicted foundation response, and 4) a projection of the results from the measured loading to extreme loading used in assessments.

STRUCTURAL MEASUREMENTS

Measurement Plan

The goal of this measurement program was to resolve uncertainties that have significant impact on jack-up assessments. In keeping with this goal, the program was tailored to measure leg foundation fixity and wave-induced dynamic amplification. Leg

References and illustrations at end of paper.

fixity was selected because a change in its modeling, from "pinned bottom" to "rigidly-fixed bottom", reduces the peak leg bending moment, under a given load, by a factor of two. The same change also reduces the natural period of the structure by a factor of two, causing a significant reduction in the dynamic load amplification. Dynamic amplification was selected because it can also have a significant impact on the assessment of a rig's maximum water depth capability, as discussed in Reference 1.

Following the establishment of the specific program objectives, the KEYES 301 was chosen for the measurements. This rig (described in Table 1) was under contract to work in relatively deep water where it would likely exhibit measurable dynamic amplification. In addition, the rig would be operating during the winter, when moderately large storms would likely provide large response of the structure.

During the measured storm, the rig was located at Eugene Island Block 295 in 215 feet of water. During pre-loading, the legs penetrated 51 feet into the soft clay bottom. Soil at this site is primarily a highly plastic, normally consolidated clay to a depth of 141 feet below the mudline. A reasonably continuous layer of medium dense, silty fine sand extends from 52 feet to 67 feet below the mudline. Figure 1 shows the soil profile for this site, along with the approximate position of the embedded jack-up leg. As shown, the spud can and the lower structural bays penetrated the soil until the tip of the spud can was close to the top of the sand layer. Table 1 contains additional description of the soil.

Measurement System

A measurement system was designed to measure the following quantities: 1) reaction loads at two elevations on one of the three legs, 2) motion of the hull, and 3) metocean conditions.

Leg bending and axial loads were measured with strain gages placed on vertical chord members on the bow leg. The two gaged elevations were 1) 100 feet above the tip of the spud can (49 feet above the mudline), and 2) 315 feet above the tip of the spud can (49 feet above mean sea level), as shown in Figure 2. At each elevation, a pair of strain gages was installed on each of the three chords and configured to measure axial force in the chord. (The gages were located where chord bending was expected to be negligible, thereby accommodating the potential loss of a gage.)

Additional sensors were located on the deck. Accelerometers provided hull motions data, suitable for characterizing the bending and torsional natural periods. Wave height and wind speed sensors provided metocean data for use in estimating the load history in subsequent time-domain simulations. Figure 3 shows a layout of the equipment as it was installed.

A 16-channel data acquisition system sampled the sensor outputs at a rate of three samples per second. The data acquisition equipment included Wheatstone bridge circuitry for the strain gages, electronic filters and amplifiers to condition the analog signals, and a PC that could be remotely-controlled via phone modem from shore. The data were recorded by the PC under "threshold" control (only during large seastates to limit the volume of data recorded) and logged to high-capacity optical disks.

The installation of the measurement system was carried out in two phases: one in a shipyard and the second on site. In the shipyard, subsea strain gages were placed on vertical chord members at an elevation near the anticipated mudline on the bow leg. The remaining sensors were installed and the data acquisition system was connected on site.

Measured Results

A one-hour segment of the most severe storm data was processed to determine the dynamic response of the rig. Using spectral analysis techniques (See References 2 and 3), the fundamental natural periods, mode shapes, and damping ratios were determined from hull acceleration spectra. Spectral processing of the leg strain measurements provided the dynamic amplification, leg loads, and foundation fixity. Table 2 shows a summary of the measured quantities. The procedure for inferring the leg loads and the foundation fixity is described below.

Conversion of Strains to Leg Loads

The leg bending and axial loads were inferred from the chord member strains using a composite beam model of the leg. This model assumes that these loads are carried entirely by axial force in the chords. For each time sampled, the strains were converted to loads using the following equations:

$$F_z = AE(\epsilon_1 + \epsilon_2 + \epsilon_3) \quad \text{(1)}$$

$$M_y = (1/2)AEd(\epsilon_3 - \epsilon_3) \quad \text{(2)}$$

$$M_x = (\sqrt{3}/6)AEd(2\epsilon_1 - \epsilon_2 - \epsilon_3) \quad \text{(3)}$$

In these equations, ϵ_i is the axial strain in each of the three chords, A is the cross-sectional area of a chord, E is the elastic modulus of steel, and d is the distance between chord centers. Figure 4 shows a cross-section of the jack-up leg and defines the loads and strains used in the composite beam model.

Measured Fixity

The foundation fixity was quantified by using the location of the inflection point (elevation of zero bending) on the leg bending moment profile. For example, with a concentrated, horizontal load applied at the hull, zero fixity results in an inflection point at the tip of the spud can. Under the same load, a leg rigidly fixed in the soil would have an inflection point near mid-length, depending on the leg/hull flexibility.

The leg bending measurements yield two points on a bending moment profile. To determine the inflection point, the shape of the bending moment profile is estimated. Under distributed wave loading, the leg's bending moment profile, and thus its inflection point location, is highly uncertain. To avoid this uncertainty, a method was developed to calculate the inflection point based on the inertial loads.

The calculation method utilizes several properties of the jack-up's structural response. First, the inertial (resonant) response is separable from the quasi-static wave response under the measured conditions, as demonstrated by the spectrum in Figure 5. Second, because of the large mass of the hull, the inertial load can be represented by a concentrated force acting at the hull's center of mass. Under this inertial load, the leg bending moment profile is linear.

The inflection point based on the inertial loads is calculated as follows. First, the leg bending moment data from the two gaged elevations are filtered to preserve only the frequency range which contains the inertial response. Second, the extreme values of the two sets of inertial bending moment data are determined at consistent times. Third, the two bending moment values, with sign preserved, are plotted versus elevation. Finally, a straight-line bending moment profile is drawn between the points and the elevation of the inflection point is determined. (Note: For these measurements, the standard deviations of the two sets of bending data were also evaluated and found to give the same inflection point as inferred from the maxima.)

Total Load Estimate

Due to soil nonlinearities, the inflection point was associated with the load level encountered during the measurements. This load level was calculated based on the combination of the quasi-static wave load and the inertial load during the extreme event. Equations to determine the loads assume a bow-to-stern loading direction (consistent with the storm direction observed) and also assume that all three legs have the same bending. (In general, the legs will have different bending moments due to the phase of the wave loading; however, this estimate is reasonable when inertial loads are large.)

The total overturning moment (OTM) and the equivalent shear force (F_{eq}) on one leg (concentrated at the hull) were estimated using the following equations:

$$OTM = F_z D + 3M_x \quad (4)$$

$$F_{eq} = OTM / 3L \quad (5)$$

In these equations, F_z is the axial force measured in the leg; M_x is the leg moment measured at the mudline; D is the distance from the center of the bow leg to the line through the centers of the stern legs; and L is the distance from the mudline to the main deck.

For the measured storm, the leg foundation fixity was determined using the procedure described above. The inflection point was found to be 186 feet above the tip of the spud can and the associated total equivalent force was 41 kips.

As will be discussed in later sections, higher loads tend to reduce the stiffness of the leg foundation, leading to a lower inflection point and a longer structural natural period. However, as will be shown in the next section, the estimated 41-kip load during the storm is large enough to initiate some soil nonlinearity. While not required, this enhances the value of the measured data for calibrating the leg foundation model.

LEG FOUNDATION MODEL

Prior to the measurements, a jack-up leg foundation model was developed to account for 1) the lateral resistance afforded by any leg members embedded in the soil and 2) the spud can's lateral and rotational resistance. Shown in Figure 6, the leg foundation model incorporates discrete nonlinear lateral springs along the penetrated portion of the leg and a nonlinear rotational spring located 6 feet above the tip of the spud can. A simplified model of the leg-to-hull connection provides rigid rotational restraint midway between the leg guides.

The model was constructed in a relatively crude form, due to the highly uncertain conditions of the soil surrounding the leg and spud can. In this model, a very qualitative treatment was given to effects such as soil remolding, sand layer influence, and leg/soil contact.

Fixity from Embedded Leg Members

The lateral soil resistance was modeled based on concepts proposed by Matlock (see Ref. 4) and the equivalent leg foundation properties are listed in Table 1. The diameters of the individual members (i.e., leg chords and braces) were assumed to be the appropriate characteristic dimensions for developing "p-y", or load-deflection, curves. This is considered a valid assumption because of the open structure of the leg. The p-y curves for each member were summed to form a composite p-y curve for each bay. Only one face of each leg was assumed to be in contact with the soil and thus contribute to lateral resistance. The actual leg-soil contact depends on a number of factors, such as the degree of hole collapse, but the assumption here is believed to be reasonable.

Fixity from Spud Can

Foundation fixity due to spud can rotational resistance was modeled as a linearly elastic-perfectly plastic rotational spring. For elastic loading conditions, the rotational stiffness is taken to be that of a circular disk in an elastic half space,

$$K_r = 16GR^3 / 3(1-\nu) \quad (6)$$

where G is the soil shear modulus, R is the spud can radius, and Poisson's ratio (ν) is assumed to be 0.5.

For initial stiffness at small strains, the appropriate shear modulus may be on the order of 1000 times the undrained shear strength. However, for the measured conditions the estimated load levels were undoubtedly well into the nonlinear range, as will be demonstrated below.

The ultimate moment capacity of the disk was developed using plastic limit analysis, assuming a spherical failure surface centered on the disk. Because of the weak material above the spud can, only one-half of this failure surface was assumed to be effective. The resulting equation for ultimate moment capacity is

$$M_u = (1/2)\pi^2 R^3 S_u \qquad (7)$$

where S_u is the average undrained shear strength over the failure surface. The shear strength was assumed to be approximately 660 psf, based on the soil profile in Figure 1 where the disk is effectively embedded 45 feet below the mudline. The average value of the spud can radius was assumed to be 20 feet. Thus, the ultimate moment capacity used in the model was 2.6 X 10^7 ft-lbs.

Assuming a rigid rotational restraint at the spud can, the measured load of 41 kips applied at the hull would develop a moment at the spud can on the order of 7.2 X 10^6 ft-lbs. This is about one-third of the ultimate moment capacity of the spud can (ignoring lateral support) and is thus well above the load level which would initiate soil nonlinearity. At the measured load level, the shear modulus might be an order of magnitude less than the peak value associated with small strains. Thus, the rotational restraint is linearized using a shear modulus equal to 100 times the undrained shear strength. Based on Equation 6 and using this shear modulus associated with the measured load level, the initial stiffness used in the nonlinear relationship was 5.7 X 10^9 ft-lbs/rad.

Because of the very approximate nature of the initial stiffness and the ultimate moment capacity, the effects of large variations in these parameters are discussed in a later section. No explicit attempt was made to account for the effect of the underlying sand layer, except in a parametric sense.

MEASURED VERSUS PREDICTED FIXITY

A prediction of fixity was developed based on the leg foundation model and compared with that inferred from the measurements. For consistency with the measurements, a 41-kip load was applied to the model and the leg bending moment profile was calculated. As shown in Figure 7 (Case 1), this bending moment profile has an inflection point located 158 feet above the tip of the spud can. However, this estimate is based on the assumption that the leg/hull connection is rigid. In order to obtain a predicted fixity that was directly comparable to that measured, a correction term was developed.

The correction for leg/hull flexibility was developed using results from the jack-up structural model. This model includes details of the leg/hull connection such as restraint from leg guides and vertical stiffness from pinions (or rack chocks, if engaged). The stiffness of the leg/hull model was calibrated using the measurements of natural period and leg fixity. Assuming a rigid leg/hull connection, as modeled in the leg foundation model, the case of rigid soil has an inflection point half-way between the mudline and main deck of the hull (see Figure 7). Based on the more detailed leg/hull model, the same rigid soil case has an inflection point that is 30 feet higher due to leg/hull flexibility. Thus, the correction that was added to the predicted inflection points is

$$L_c = (L_p/L_{rigid})X(30\text{ feet}) \qquad (8)$$

where L_p is the uncorrected prediction of the inflection point and L_{rigid} is the distance from the tip of the spud can to the point half-way between the mudline and the main deck.

After this correction, the predicted inflection point was 181 feet above the tip of the spud can. This compares favorably with the 186 feet inferred from the measurements. Considering the expected foundation variability, the match is likely fortuitous to some degree. However, the measurements and predictions both indicate that the foundation provides very significant restraint compared to the typical "pinned" spud can model.

Uncertainties in the Predicted Fixity

Under the measured load of 41 kips, the predicted fixity was relatively insensitive to changes due to uncertainties in the assumed initial soil stiffness and the ultimate soil resistance.

Figure 7 (Cases 2 and 3 for the 41-kip load) shows the effects of increased initial stiffness. The inflection point moved up about 12 feet due to the combined effect of increased initial stiffnesses by 300 times for the lateral springs (Case 2) and by an order of magnitude for the spud can tip springs (Case 3). The initial stiffness reflects how closely the model approximates the cube root function of the soil's p-y curve, which has infinite slope at very small loads. The choice of initial stiffness is quite arbitrary (not based on measurements) and the proper value can vary by orders of magnitude, depending on the load range of interest.

Figure 7 (Case 4 for the 41-kip load) shows the effect of increased ultimate resistance. The inflection point moved up an additional 5 feet when a 20 percent increase in ultimate resistance was coupled with the increased initial stiffnesses. The ultimate resistance of the soil can usually be estimated reasonably accurately; estimates should certainly not be off by orders of magnitude. The 20 percent increase was chosen arbitrarily to demonstrate the effects. The actual uncertainty in this analysis could be larger due to the assumption that the soil resistance acts only on one face of the leg, as well as the existence of the sand layer at the tip of the spud can. However, under the

41-kip load, further increases in ultimate resistance should have a negligible effect because they have not been fully mobilized at this loading level.

EXTRAPOLATION TO EXTREME LOADING

After the leg foundation model was calibrated at the measured loading, it was used to predict leg foundation fixity under an extreme load more typical of that used in assessments. As load level increases, the leg foundation fixity reduces markedly due to the high degree of nonlinearity in the soil. However, even under an extreme load, the leg foundation fixity model provides considerably more structural resistance than a "pinned" spud can model.

As shown in Figure 7 (Case 1 - 500-kip load), the predicted bending profile has an inflection point located 95 feet above the tip of the spud can. After the leg/hull flexibility correction, this value increases to 109 feet.

Figure 7 (Cases 2, 3, and 4 - 500-kip load) shows the effects of uncertainties in the soil parameters for the extreme load. As shown, an increased initial stiffness of the lateral p-y curves (Case 2) has a negligible effect on the location of the inflection point. An increased initial stiffness of the tip rotational spring (Case 3) likewise has a negligible effect. An increase in ultimate resistance (Case 4) has the most pronounced effect; further increases in this value would likely provide substantial increases in the leg foundation fixity.

A jack-up structural analysis which incorporates the leg foundation model results in a 20 percent reduction in both structural natural period and peak leg moment when compared to an analysis using a pinned spud can model. For purposes of this comparison, the pinned model was located at the tip of the spud can. In this analysis, with a wave period associated with extreme seastates, a 20 percent drop in natural period leads to 20 percent less total wave-induced load. This is due to the decrease in wave-induced dynamic amplification. Under a given load, the peak leg bending moment just under the hull, where member stress is typically the largest, is cut by an additional 20 percent.

CONCLUSIONS

A jack-up leg foundation model has been calibrated based on measurements conducted on a jack-up operating at a site with significant leg penetrations in soft clay. This model is appropriate for site-specific jack-up structural assessments where soils data are available. Application of this improved model has been shown to significantly reduce overconservatisms in structural assessments when compared to the commonly-used "pinned" spud can model.

Future research is needed in the area of jack-up structural assessments, including additional measurements with accompanying foundation calibration. Jack-up measurements are needed to calibrate total structural loads, including quasi-static wave load and wave-induced dynamic amplification. Measurements are particularly needed at a site with a firm, sandy bottom. · Smaller-scale experiments could shed light on ultimate resistance of jack-up leg foundations. Analytical work could result in methods for estimating hysteretic soil damping under extreme loading; this would provide additional damping (compared to what has been measured) and further reduce the predicted dynamic amplification.

ACKNOWLEDGMENTS

The authors wish to acknowledge the following for their contributions: Norman Peterson of Electronics Measurement Co. for his field installation of the strain gages; Fred Spence of Triton Systems for his field technical support; Keyes Offshore, Inc. for their cooperation during the field measurements; Exxon USA for their logistics and planning support; Penrod Drilling for their cooperation during pilot measurements on the PENROD 63; and Friede and Goldman for providing rig drawings and information on key design details.

REFERENCES

1. Grenda, K. G., "Wave Dynamics of Jackup Rigs", Proceedings, Offshore Technology Conference (OTC 5304), May, 1986.

2. Bendat, J. S., and Piersol, A. G., Random Data: Analysis and Measurement Procedures, published by John Wiley & Sons, Inc., 1971.

3. Campbell, R. B., and Vandiver, J. K., "The Determination of Modal Damping Ratios From Maximum Entropy Spectral Estimates", Journal of Dynamic Systems Measurement, and Control, ASME, March 1982, Vol. 104.

4. Matlock, H., "Correlations for Design of Laterally Loaded Piles in Soft Clay," Proceedings Offshore Technology Conference (OTC 1204), April, 1970.

5. Hardin, B. O., and Drnevich, V. P., "Shear Modulus and Damping in Soils: Design Equations and Curves," ASCE, Journal of the Soil Mechanics and Foundations Division, Vol. 98, No. SM7, pp. 667-692, July, 1972.

6. Odland, J., "Response and Strength Analysis of Jack Up Platforms," Norwegian Maritime Research Report, No. 4, 1982.

7. Poulos, H. G., and Davis, E. H., Elastic Solutions for Soil and Rock Mechanics, John Wiley and Sons, 1974.

8. Rapoport, V., and Young, A. G., "Foundation Performance of Jack Up Drilling Units, Analysis of Case Histories," Proceedings, Int. Conference on Mobile Offshore Structures, London, 1987.

Table 1 - Description of Jack-Up and Leg Foundation

- KEYES 301 (owned by Keyes Offshore, Inc.)
- L-780 Mod II Rig Type (designed by Friede and Goldman)
- Three Independent Legs with Three Chords per Leg
- Site: Eugene Island 295A (215 feet of water)
- Air Gap of Elevated Hull = 65 feet
- Leg Foundation Model Values for Analysis:
    - Leg Stiffness, EI (lb-ft^2) = 2.15 X 10^{12}
    - Spud Can Diameter = 39.5 feet
    - Penetration of Tip of Spud Can = 51 feet
    - Depth of Rot. Soil Spring Model = 45 feet
    - Equiv. Leg Length Above Mudline = 305 feet
    - Characteristic Member Diameter = 0.89 feet
    - Length of Structural Bays = 12.0 feet
    - Effective No. of Members/Bay = 10.8
    - Soil Strength at Mudline = 100 psf
    - Soil Strength Gradient = 12.5 psf/ft
    - Soil Submerged Weight = 50 pcf

Table 2 - Summary of Measured Quantities

- Maximum Wave Height (crest to trough) = 25 feet
- Structural Natural Period (bending) = 4.7 sec.
- Structural Natural Period (torsion) = 3.9 sec.
- Damping Ratio (percent of critical) = 2.0% (+/- 1.0%)
- Leg Fixity (infl. pt. to spud can tip) = 186 ft
- Estimated Total Overturning Moment = 37,600 ft-kips
- Est. Equivalent Load at Top of Leg = 41.1 kips/leg

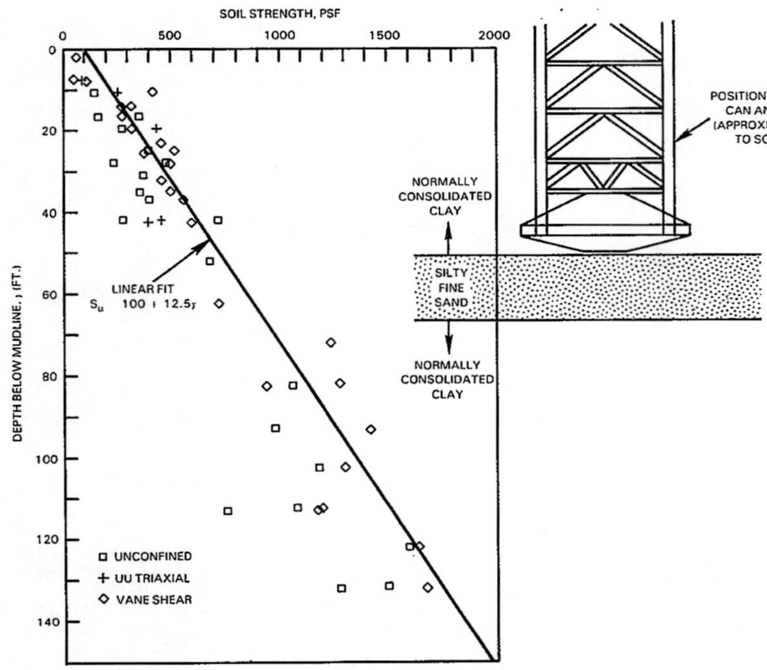

Figure 1. Soil strength profile

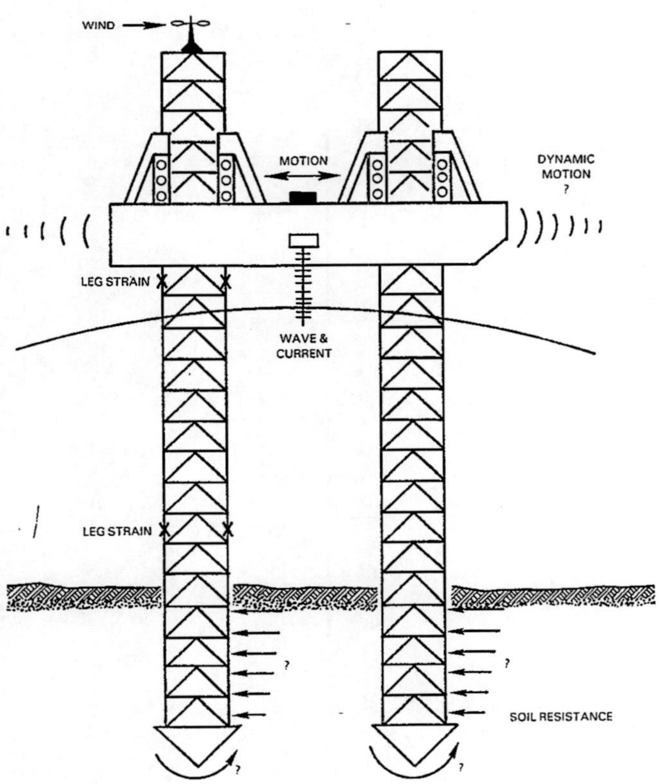

Figure 2. Profile of measurement system

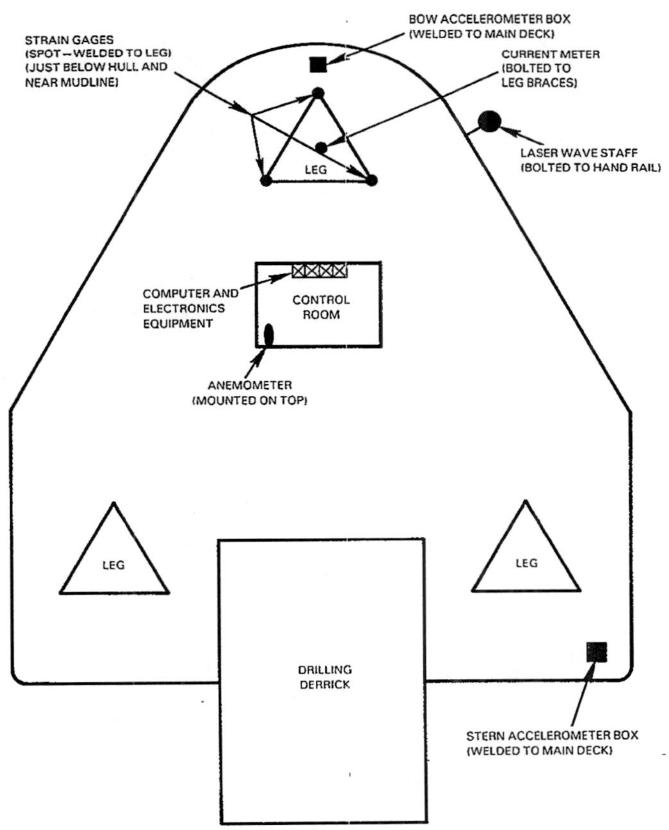

Figure 3. Layout of measurement equipment

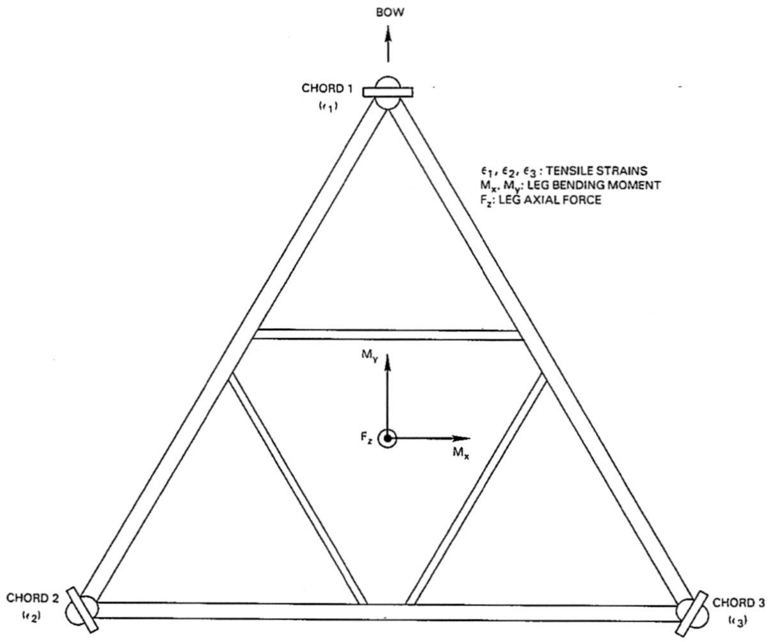

Figure 4. Cross-section of jack-up leg

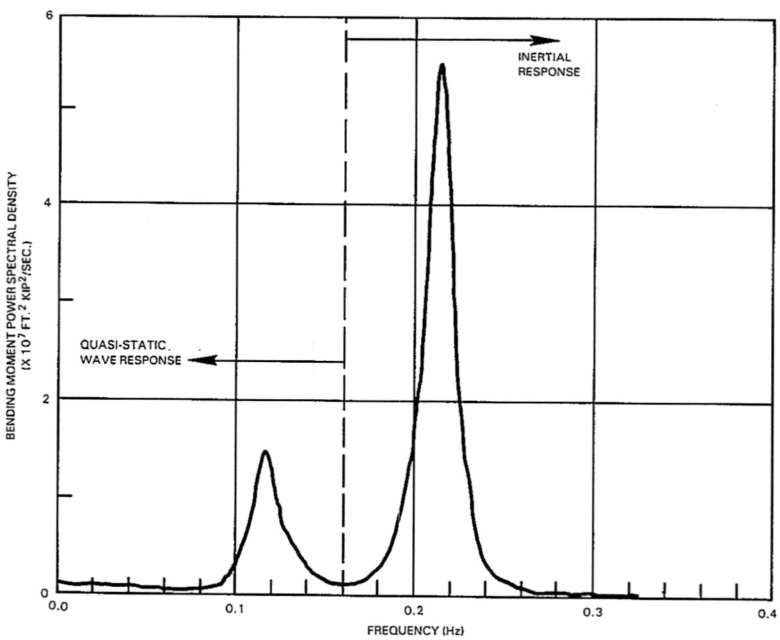

Figure 5. Measured spectrum of leg bending moment

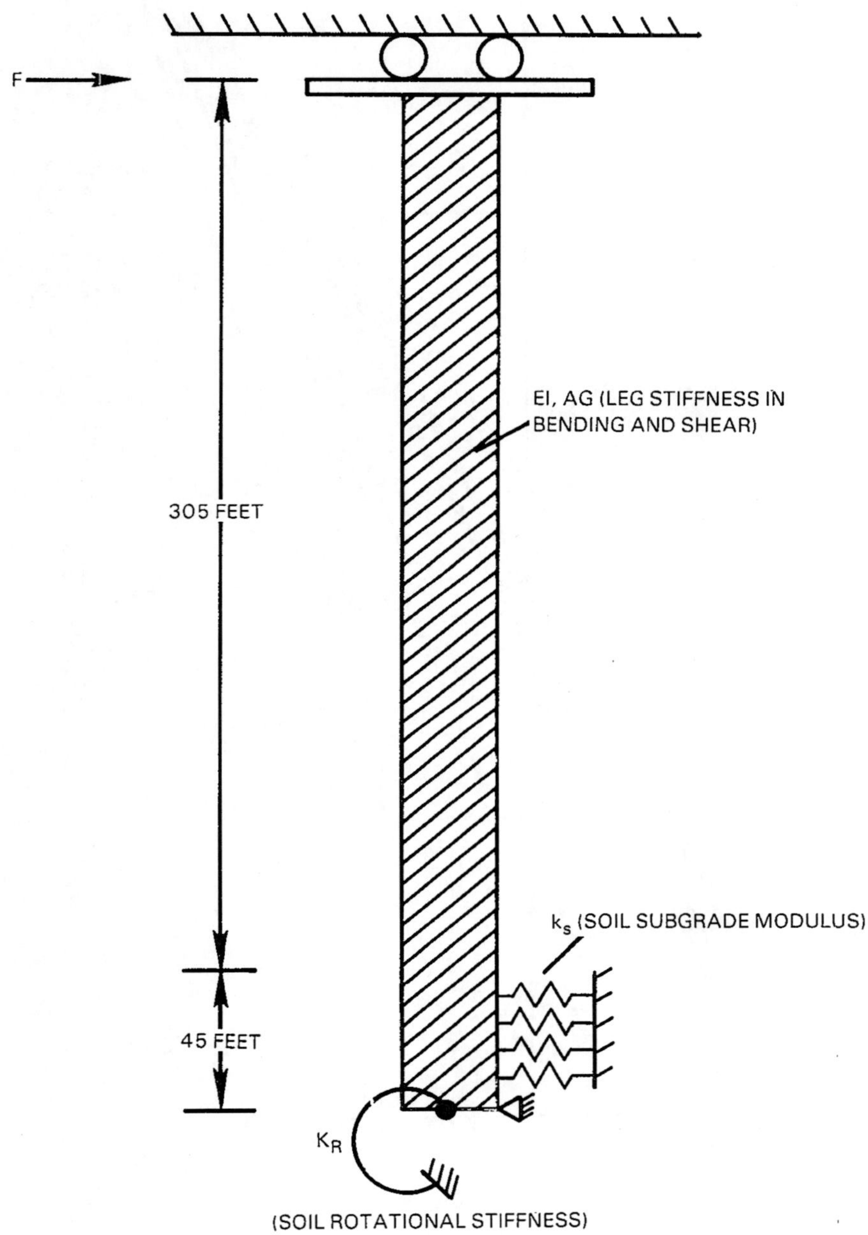

Figure 6. Leg foundation model

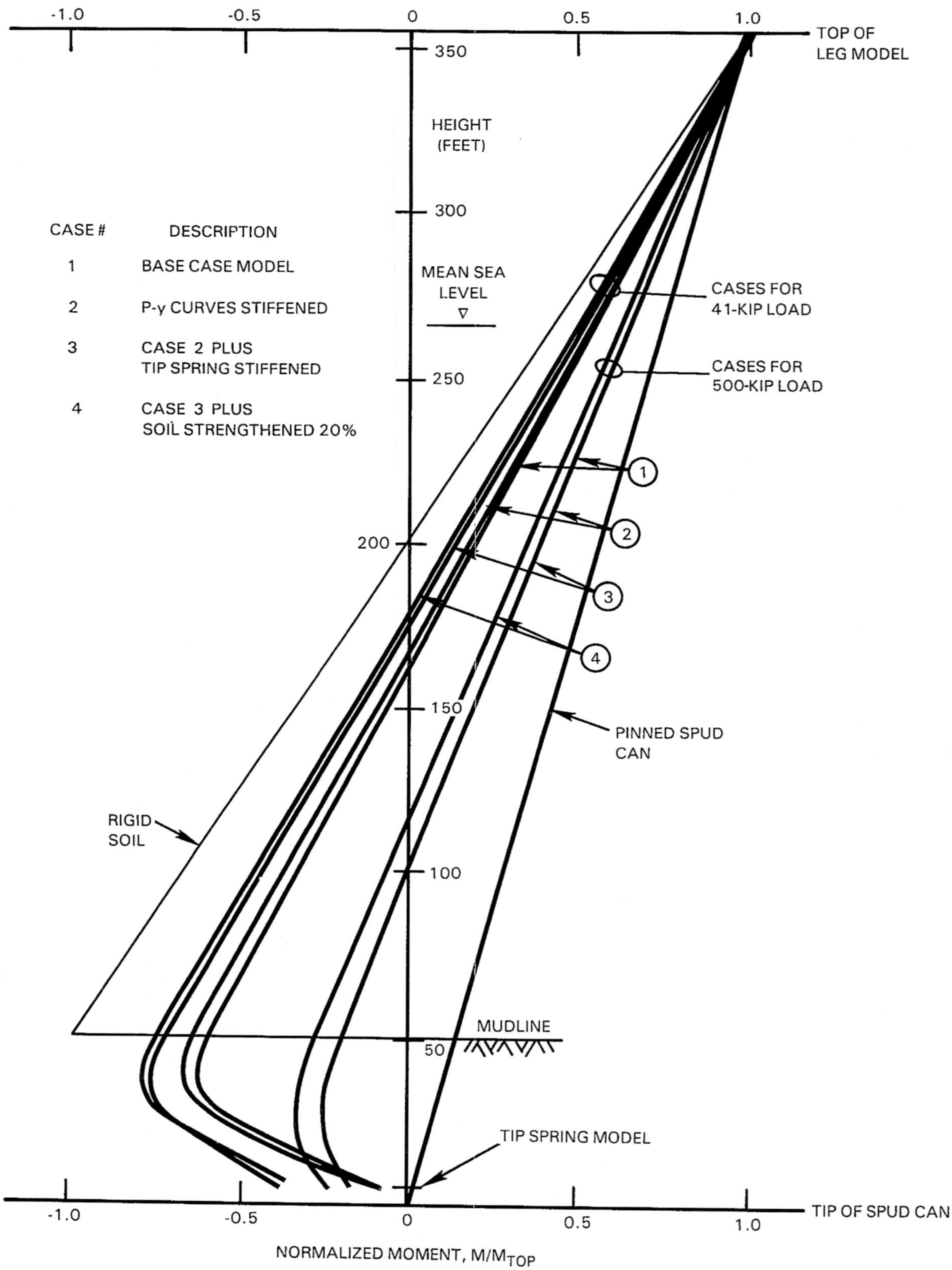

Figure 7. Predictions of leg bending moment

OTC 6335

Design of Submarine Pipelines Against Upheaval Buckling

A.C. Palmer, Andrew Palmer & Assocs.; C.P. Ellinas, Advanced Mechanics & Engineering;
D.M. Richards, Lloyds Register; and J. Guijt, A/S Norske Shell

This paper was presented at the 22nd Annual OTC in Houston, Texas, May 7–10, 1990.

ABSTRACT

This paper describes part of a comprehensive joint-industry project on upheaval buckling. It develops a semi-empirical simplified design method and detailed design methods based on a new numerical analysis, and illustrates their application by examples. It assesses alternative design strategies, and the implications of strain-based design.

INTRODUCTION

When a pipeline is operated at a temperature and pressure higher than ambient, it will try to expand. If the line is not free to expand, the pipe will develop an axial compressive force. If the force exerted by the pipe on the soil exceeds the vertical restraint against uplift movement created by the pipe's submerged weight, its bending stiffness, and the resistance of the soil cover, the pipe will tend to move upward, and considerable vertical displacements may occur. The pipeline response might then be unacceptable because of excessive vertical displacement or excessive plastic yield deformation. Upheaval buckling is hence a failure mode that has to be taken into account in the design of trenched and buried pipelines.

The possibility of upheaval buckling is of increasing concern to the operators of flowlines in the North Sea and elsewhere. Their concern has been heightened by upheaval buckling incidents in the Danish and Norwegian sectors, and by the cost of protective measures such as rock dumping. Shell International Petroleum Maatschappij (SIPM) initiated a comprehensive upheaval buckling research program in 1987. In its final form, the research program included six phases:

1. review of available models;
2. comparative analysis;
3. cover response, testing and modelling;
4. numerical upheaval buckling model;
5. analysis of alternative concepts;
6. development of an upheaval buckling guideline.

The present paper is one of a series describing the results of the programme. It concentrates on phases 5 and 6, which were subcontracted by SIPM to a joint venture of three UK engineering companies, Andrew Palmer and Associates, Advanced Mechanics and Engineering, and Lloyds Register.

The paper is primarily concerned with the design aspects of upward buckling of buried pipelines. Unburied pipelines can buckle in a related but different mode, in which they snake sideways across the seabed. This mode too is important in practice, and can play a major part in relieving the effects of axial force, but it is outside the scope of the present paper.

The paper first briefly describes the phenomenon, and puts forward a simplified method for preliminary design. It goes on to describe a conceptual design method based on the application of the UPBUCK

References and figures at end of paper

program, and applies the methods to a design example. The paper then considers alternative strategies to resolve upheaval problems, and examines in more detail one of these strategies, a reduction in wall thickness made possible by the replacement of stress criteria on longitudinal stress with strain-based design criteria.

UPHEAVAL BUCKLING

The source of upheaval buckling is an interaction between the axial compressive force and overbend "hill" imperfections in the pipeline profile.

Figure 1 illustrates a sequence of events which initiates buckling in a buried pipeline. The pipeline is laid across an uneven seabed (Figure 1a), and later trenched and buried (1b). The trenching and burial operations modify the profile of the foundation on which the pipe is resting, so that it is not precisely the same as the original profile. Trenching may smooth the profile overbends, but may also introduce additional imperfections, if, for instance, a lump of bottom soil falls under the pipe.

When the pipe goes into operation, its internal pressure and temperature are higher than when it was installed and trenched, and the axial force becomes compressive. The effective axial force in a constrained pipeline has two components, both of which contribute towards buckling (Palmer [1]). The axial force in the wall is the resultant of a compressive constrained thermal expansion component and a tensile Poisson component, while in addition there is a compressive force component in the contained fluid.

On an overbend, the axial compressive force reduces the upward reaction between the foundation and the pipeline (Figure 1c). A further increase of operating temperature and pressure may reduce the reaction to zero. The pipe then lifts on the overbend, moves towards the surface of the cover, and may break out through the surface (Figure 1d).

SIMPLIFIED ANALYTICAL MODEL

The stability of the pipeline in its initial position turns out to depend on the local profile of the pipe in contact with its foundation, and on whether or not enough downward force is present to hold the pipe in position. If the pipe does not move, the governing factors are the constrained axial force and the pipe flexural rigidity, and other parameters have no effect.

An arbitrary pipeline profile can be defined by a height y and a horizontal distance x. The pipeline is idealised as an elastic beam which carries an axial compressive force P and has flexural rigidity EI. It follows from elementary beam-column theory that the downward force w(x) per unit length required to maintain the pipeline in equilibrium in this position is

$$w(x) = -EI\, d^4y/dx^4 - P\, d^2y/dx^2 \quad \ldots(1)$$

and therefore depends on the profile shape through the fourth and second derivatives.

Consider first a simple sinusoidal profile imperfection of height δ and length L defined by

$$y = \delta \cos^2(\pi x/L) \text{ in } -\tfrac{1}{2}L<x<\tfrac{1}{2}L \quad \ldots(2)$$

The downward force required to maintain this profile is therefore

$$w(x) = (-8\delta EI(\pi/L)^4 + 2\delta P(\pi/L)^2)\cos(2\pi x/L) \quad \ldots(3)$$

and has its largest numerical value at the crest of the imperfection, where

$$w = 2\delta P(\pi/L)^2 - 8\delta EI(\pi/L)^4 \quad \ldots\ldots(4)$$

the downward force per unit length required to stabilise the pipeline at the crest of the profile imperfection.

Equation (4) applies to a particular profile shape, but the specific shape only affects the coefficients and not the general form of the equation. It can be rewritten

$$\Phi_w = 2\pi^2\Phi_L^{-2} - 8\pi^4\Phi_L^{-4} \quad \ldots\ldots\ldots\ldots\ldots(5)$$

a relationship between a dimensionless maximum download parameter Φ_w defined by

$$\Phi_w = wEI/\delta P^2 \quad \ldots\ldots\ldots\ldots\ldots\ldots\ldots(6)$$

and a dimensionless imperfection length Φ_L defined by

$$\Phi_L = L(P/EI)^{1/2} \quad \ldots\ldots\ldots\ldots\ldots\ldots\ldots(7)$$

This suggests that a useful way of representing the results of upheaval buckling calculations is to plot Φ_w against

Φ_L. The resulting plot is universal, and forms a valuable summary, which can include both the results of numerical calculations and then observations from full-scale field experience and tests.

All points that represent a single imperfection shape will lie on a single curve on a Φ_w against Φ_L plot. Points that represent different imperfection shapes will lie on different curves, and the variability between curves will reflect the effect of imperfection shape (the hardest parameter to determine reliably in practice). The form of equation 5 suggests that the functional relationship between Φ_w and Φ_L is

$$\Phi_w = c\Phi_L^{-4} + d\Phi_L^{-2} \quad \ldots\ldots\ldots\ldots\ldots\ldots (8)$$

where c and d are constants to be determined numerically. This is done by plotting $\Phi_w\Phi_L^2$ against Φ_L^{-2}, in Figure 2, on which the points represent the results of numerical calculations with the UPBUCK program described below. The rightmost group of points represents imperfection profiles whose shape is an upheaval foundation, that is, a foundation in the shape of a pipeline supported by axial force in a post-upheaval mode. The remaining points correspond to "prop" imperfection profiles, whose shape is that taken up by a pipeline laid across a single isolated hill in a horizontal profile. The final relationship turns out to be bilinear, with one pair of values c and d corresponding to small values of Φ_L and a second pair to large values.

An additional condition occurs when the profile includes a very short imperfection, so that the pipeline is only in contact with the crest of the imperfection. The length Φ_L is then immaterial, and this condition forms a cut-off for short imperfections.

The results are combined in Figure 3, a universal curve which can be used directly for design. The three conditions that together give the design download required for stability are:

$$\Phi_L<4.49, \quad \Phi_w = 0.0646 \quad \ldots\ldots\ldots\ldots (9)$$
$$4.49<\Phi_L<8.06, \quad \Phi_w = 5.68/\Phi_L^2 - 88.35/\Phi_L^4 \quad \ldots (10)$$
$$\Phi_L>8.06, \quad \Phi_w = 9.6/\Phi_L^2 - 343/\Phi_L^4 \quad \ldots (11)$$

In most preliminary design contexts, the designer can determine the maximum height of a profile imperfection, but not its length. An imperfection length can however be estimated from an assumption that the pipeline takes up a form dictated by the interaction of its flexural stiffness and its weight in the installed condition (not in general the same as the weight in the operating condition). Substituting this into the third equation (11), we arrive at a design formula for the required download for stability in the operating condition:

$$w = [1.16 - 4.76(EIw_0/\delta)^{1/2}/P]P(\delta w_0/EI)^{1/2} \quad \ldots (12)$$

where

w_0 is the installation submerged weight;
EI is the flexural rigidity;
δ is the imperfection height;
P is the effective axial force in operation.

This equation is used to derive the required value for download for preliminary design. Preliminary design calculations apply a spreadsheet, which compares the required download determined from equation (12) with the actual load that can be mobilised, the sum of the pipeline's submerged weight and the uplift resistance of the cover. The research carried out into uplift resistance in the course of this research programme is discussed in detail in a separate paper[2]. In brief, the model recommended for design of buried pipelines is

for cohesionless sand, silt and rock

$$q = HD(1 + fH/D) \quad \ldots\ldots\ldots\ldots\ldots (13)$$

for cohesive clay and silt cover

$$q = cD \min[3, H/D] \quad \ldots\ldots\ldots\ldots\ldots (14)$$

where

q is the uplift resistance per unit length of pipe;
H is the cover depth (from the top of the pipe to the surface);
D is the outside diameter of the pipe;
is the submerged unit weight of the cover material;
c is the shear strength; and
f is an uplift coefficient, determined experimentally and taken as 0.5 for dense materials and 0.1 for loose materials.

The reader is warned that this simplified method is semi-empirical, and that on its own it will not normally be adequate for design. It has been calibrated against UPBUCK, but does not always yield conservative results, especially if there is a possibility of plastic deformation of the pipe wall.

CONCEPTUAL AND DETAILED DESIGN

Conceptual and detailed design are based on the UPBUCK finite-element program described in a separate paper[3]. The program takes account of a number of factors not included in the simplified analytical procedure. These factors have a significant effect on the response of the pipeline after it has begun to move in response to axial force, and thence on its post buckling behaviour and the amplitude of the final displacement. Among the factors are:

1 the finite axial stiffness of the line, which determines how rapidly the axial force diminishes as the line moves upwards;

2 the pipeline's longitudinal resistance to movement through the soil, which determines how far the pipeline can slide towards a developing buckle;

3 the incremental flexural rigidity for large deflections, which determines whether the pipeline's resistance to bending decreases as large displacements occur: the rigidity may be influenced by plastic bending deformation.

4 the finite stiffness of the foundation, which allows the foundation to deflect under forces applied by the pipe, and slightly modifies the deflected form of the pipe.

UPBUCK makes it possible for the user to calculate the response of the pipeline to increasing operating pressure and temperature, and to follow the deformation into the post-buckling range.

DESIGN EXAMPLE

The example is a 10-inch oil pipeline, with a maximum operating pressure of 15 MPa (2175 psi) and a maximum operating temperature of 80 deg C. The chosen steel grade is API 5L;X60, which corresponds to a specified minimum yield stress (SMYS) of 413.7 N/mm². The external anti-corrosion coating is 2 mm of polyethylene. The line is to be designed to DnV 1981 rules.

The line will be trenched, in order to protect it against fishing gear. A review of existing survey data along a nearby pipeline route has established that the seabed is relatively smooth, and that the maximum expected height of overbends will be 0.2 m. However, it is thought that trenching might induce additional imperfections in the bottom profile, and the design imperfection height is therefore taken as 0.3 m. The trench will be backfilled with rock, which has a minimum in-situ submerged unit weight of 8.5 kN/m^3 and an uplift coefficient of 0.5. The assumed in-situ submerged unit weight is a low conservative value, and corresponds to a voids ratio of 0.87 and a rock fragment density of 2650 kg/m^3. The option of trenching and rock-dumping, rather than rock-dumping alone, is selected because analysis has shown that the saving in rock cost more than outweighs the cost of trenching.

The wall thickness design arrives at a nominal wall thickness of 11.1 mm. The installation submerged weight is 591.3 N/m (60.2 kg/m, 40.5 lb/ft), and the operating submerged weight is 984.2 N/m.

The first step is preliminary design. It calculates the factor of safety on pipeline weight and cover uplift resistance taken together. This is the ratio between the required value, calculated from equation (12), and the actual value, the sum of the operating submerged weight and the cover uplift resistance taken from equation (13).

Figure 4 plots the calculated factor of safety against the depth of cover. It can be seen that if the cover is 0.6 m or less, the factor of safety is less than 1, that if the cover is 0.7 m the factor of safety is 1.11, and that if the cover is 0.8 m the factor is 1.32. This suggests that the minimum cover should be at least 0.8 m, since a 0.7 m cover leaves only a narrow margin. Accordingly, a minimum cover of 0.8 m is taken as the starting point for UPBUCK calculations.

A separate calculation determines the extreme stresses in operation, at the imperfection and under the axial loads induced by the operating temperature and pressure. Although the hoop stress is relatively low (183 N/mm², 44 per cent of yield), the longitudinal stress is high, and the von Mises equivalent stress is over yield. This condition is allowed by many codes, since the pipeline is continuously supported, but the high level of equivalent stress is an important warning that the pipeline response to axial loading may be influenced by plasticity, and that it will be prudent to carry out detailed UPBUCK calculations using the plasticity option.

Moving now to conceptual design, a series of UPBUCK calculations are carried out. In this phase of the analysis, the pipeline is treated as elastic.

The results of three UPBUCK calculations are summarised in Figure 5, which plots movement against temperature increase for cover depths of 0.7, 0.8 and 0.9 m above the peak of the 0.3 m imperfection. If the cover is 1 m (from the baseline), the pipeline begins to lift (at the top of the imperfection) at a temperature of about 70 deg C, some 10 deg C short of the operating temperature. It becomes unstable and jumps into a buckled configuration at 88.5 deg C. This option leaves too little margin between the operating temperature and the temperature at which large movements might occur, and is therefore rejected.

If the cover is 1.1 m, movements begin at about 80 deg C, and an unstable jump occurs at 95 deg C. At the design maximum operating temperature, the von Mises equivalent stress is 83 per cent of yield, which is within the maximum stress allowed by the code.

If the cover is 1.2 m, movement does not begin until the design temperature is reached. The conclusion of the conceptual design phase is therefore that a cover of 1.1 m (from the baseline) is adequate, and this value is chosen as the starting point for detailed design.

Continuing to detailed design, it has been shown earlier that the pipe is close to plastic yield, and that UPBUCK calculations using the plasticity option are required. The results show that movements initiate at almost exactly the 80 deg C design temperature, but do not exceed 10 mm until the temperature has risen to 88 deg C. The instability temperature is 97.3 deg C. Figure 6 plots the movement of the pipe at the peak of the imperfection against the temperature, and shows that instability will be accompanied by very large movements, in which the pipeline will break above the surface of the cover. However, there is an adequate temperature increase margin between the operating temperature and the instability temperature.

A cover of 1.1 m, measured from the baseline, is therefore selected. It corresponds to 0.8 m cover at the peak of the 0.3 m imperfection. It happens to coincide with the value reached by the simplified method, though this is not always the case.

This cover is only required on the overbends. If the critical overbends can be confidently identified from a survey, rock need only be dumped on those overbends. It will usually be cheaper to devote resources to survey rather than to continuous dumping. This design option was recently applied to the stabilisation of the Tern and Eider pipelines[4]. If on the other hand the survey accuracy is not sufficient to locate the critical overbends, the whole length must be covered.

ALTERNATIVE DESIGN STRATEGIES

The simplest and most straightforward way of stabilising a pipeline against upheaval is to bury it. This option is often difficult and expensive, and prompted a search for alternatives. Phase 5 of the research programme examined some forty alternative design strategies, evaluated each one, and identified the most generally attractive and applicable.

Two obvious strategies are to reduce the design operating temperature and pressure, and to increase the submerged weight of the pipeline. A reduction in operating temperature is generally impracticable, but could be accomplished by adding a heat exchanger to the system. It is not usually practicable to resolve the problem by an increase in submerged weight, because too much weight is required.

The remaining strategies can be classified into groups. The most promising techniques are listed in table 1.

The first group is based on reducing the driving force, the axial compressive force in the pipeline. The most direct method is to reduce the wall thickness of the line. This reduces the temperature component of the effective axial force, which is proportional to the wall thickness (and is usually the largest component), and leaves the pressure component almost unchanged. A reduction in wall thickness can be achieved by increasing the design factor to the highest allowable level, or by increasing the grade of steel, or by adopting strain-based design criteria, a possibility examined below in section 7.

A second method is to increase lay tension. Residual lay tension balances part of the compressive force induced by operating, and therefore reduces the resultant force. A difficulty is that residual tension cannot be measured directly, but must be calculated from the lay conditions, and that its continued presence in the line depends on there being no lateral movements. A third alternative is to preheat the line, and to allow it to move to relax compressive forces induced by preheat: the application of this technique to the Sun Oil Glamis flowlines is described in a separate paper[5].

A second group of methods depends on making a radical change in the structure of the pipeline. One alternative is to replace one or more single lines by a closed bundle supported on spacers in a carrier (Palmer[6]), or equivalently by a pipe-in-pipe system in which an internal flowline is supported in an outer pipe. The internal lines in the bundle then develop axial compressive forces in operation, but those forces can be balanced by tensile forces in the outer carrier, through end bulkheads and possibly intermediate bulkheads. The internal lines may bow laterally on the spacers, but the movements are controlled and the bending stresses induced may remain at an acceptable level. This effect can be further enhanced by making the internal bundle helical (Duxbury[7]): this geometry increases the reduction in axial compressive force that accompanies outward movements of the internal lines. A decision to opt for a bundle concept naturally has broader implications that extend far beyond upheaval buckling, which is only one of the factors involved.

A second radical alternative is to replace a rigid steel pipeline with a flexible. Flexibles are subject to upheaval buckling (Bournazel[8], Putot[9,10]), primarily because of the pressure effect. The tendency to buckle in service can be reduced by laying or trenching under internal pressure, or by modifying the internal structure away from a "balanced" design, so as to produce a pipe which tends to contract axially when loaded by internal pressure.

A third group of methods is based on stabilising the pipeline with rock, but using the rock in a more efficient way, so to gain an equally effective stabilising effect with less rock. The most effective of these methods is only to place rock on critical overbends (Locke[4]), but this does of course make it essential to be able to identify the overbends confidently, a demanding survey task in deep water, or when the critical imperfection height is small. A second method is to place a geotextile over the pipe before the rock is placed. When the pipeline begins to lift, the weight of the rock on the geotextile on either side of the pipe holds the geotextile down, and generates a tension whose downward component adds to the uplift resistance of the rock above the pipeline. At the same time, the geotextile pulls the rock inward, and increases the horizontal compressive stress in the rock above the pipeline, and that in turn increases the shear resistance across potential shear surfaces above the pipe, further increasing the uplift resistance. Against this, on the other hand, the use of a geotextile in a subsea environment will require a comprehensive investigation of its long-term stability against creep and structural deterioration.

A final alternative is to not to place the rock continuously along the whole length of the pipeline, but to place the rock in intermittent dumps. This possibility is discussed in an accompanying paper[11].

The offshore pipeline industry is only now beginning to explore the many different concepts that can be used to secure effective and economical stabilisation, and much more remains to be learned.

STRAIN-BASED DESIGN CRITERIA

In the past, the conventional approach to wall thickness design was generally governed by two stress requirements. The first requirement is that the hoop (circumferential) stress not exceed a defined fraction of the specified minimum yield stress (SMYS). A second requirement also has to be satisfied, and takes account of longitudinal stress; the maximum value of an equivalent stress must not exceed a specified fraction of the SMYS. The specified fraction for the equivalent stress is generally higher than the specified fraction for the hoop stress. The equivalent stress is defined either as a von Mises equivalent stress (proportional to the square root of the second invariant of the deviatoric stress tensor), or as the difference between the largest and smallest principal stresses.

If this approach is adopted for pipelines that operate at a high temperature, the second requirement often determines the wall thickness. This happens because the longitudinal stress in operation is large and compressive, and because the temperature component of stress is independent of the wall thickness. The need to reduce the equivalent stress below the allowable limit then forces down the allowable hoop stress, and this in turn leads to a high wall thickness.

An alternative approach is better. In the case of a continuously-supported constrained pipeline on or in the seabed, exceedance of an allowable combined stress does not in itself correspond to a limit state which in any way threatens the safety of the line. If the combined stress should reach yield, a limited amount of plastic deformation will occur, but it can be shown that the plastic components of longitudinal and hoop strain are small by comparison with the elastic component of hoop strain (provided that the first requirement still limits the hoop stress, and that the design factor is at customary level). The combined stress requirement can then be dropped, and replaced by a limit on strain. This is a more rational approach, and is allowed by several modern codes. The allowable strain is set at a high level, often 0.01 or 0.02, but this level is almost never reached, except when the pipe is deliberately plastically deformed in a construction operation such as J-tube pull. If the equivalent stress requirement is dropped and replaced by a limit on strain, the hoop stress condition determines the wall thickness. The resulting wall thickness is then often significantly reduced. Figure 7 is an example: it refers to an 8-inch flowline with a design temperature of 125 deg C and a design pressure of 27 MPa (3910 psi), and plots the hoop stress, the longitudinal stress, and the von Mises equivalent stress against the wall thickness. If the equivalent stress requirement is adopted, and the equivalent stress is limited to 0.96 of SMYS (as in the DnV rules), and the steel has an SMYS of 413.7 N/mm² (corresponding to API X60), the minimum wall thickness is 12.8 mm. If on the other hand the equivalent stress requirement on longitudinal stress is dropped, and the hoop stress is limited to 0.72 of SMYS (as in many codes), the minimum wall thickness can be reduced to 9.6 mm.

This reduction in wall thickness leads to a substantial reduction in axial force in operation, and to a corresponding reduction in minimum cover required for stability.

Adoption of strain-based criteria does however raise a number of wider questions, which tend to reduce the advantages. Figure 8 plots the stress path followed in a sequence of laying, increase in temperature, and increase in operating temperature, in a diagram whose axes are longitudinal stress and hoop stress. The ellipses represent successive yield loci for a von Mises material with an isotropic hardening rule. The heavy ellipse represents the nominal yield locus, corresponding to the 0.5 per cent yield locus, but yield begins before this locus is reached, and is represented schematically by the inner loci.

If a strain criterion is adopted, the stress point S that represents the mean longitudinal and hoop stresses during operation lies beyond the yield locus, even if the hoop stress is less than 0.72 of nominal yield. If the pipe then begins to bend, the points representing the stresses in the pipe in the "inside" and "outside" of the curve follow the paths shown in Figure 9b. On the inside of the curve, the longitudinal stress becomes more compressive, and the stress point moves to the left. This takes it across additional yield loci, and more yield occurs. On the outside, the longitudinal stress becomes more tensile, and the stress point moves inward, so that the incremental response is elastic.

It follows that the incremental bending of a pipeline designed according to strain criteria will always be accompanied by plastic deformation, however small the bending is. The effect is to generate both circumferential and longitudinal plastic strain. The incremental increase in bending moment is much smaller than it would be if the pipe were elastic, because only part of the pipe is responding elastically: the effect is to reduce the incremental flexural rigidity, which is one of the parameters that enable the pipe to resist upheaval buckling. At the same time, plastic deformations relieve some of the axial compressive force in the pipe wall, and therefore reduce the driving force.

Plasticity effects will therefore have a significant influence on upheaval buckling of pipelines designed according to strain criteria. The solution will be sensitive to the details of the stress-strain relation for the steel, particularly to strain-hardening. These effects are taken into account in analysis by the UPBUCK program.

CONCLUSION

This paper summarises design methods based on the results of an extensive program of research into upheaval buckling.

ACKNOWLEDGEMENT

The research described here was carried out as part of a joint industry program led by Shell International Petroleum Maatschappij and supported by the following companies:

Shell Internationale Petroleum Maatschappij
Maersk Olie og Gas A/S
Nederlandse Aardolie Maatschappij BV
Shell UK Exploration and Production
A/S Norske Shell
Britoil Plc (BP Exploration)
UK Department of Energy
Marathon Oil UK Limited
Elf Petroland BV
Occidental Petroleum (Caledonia)
Saga Petroleum A/S
Sun Oil Britain Limited
Total Oil Marine plc

The authors wish to record their gratitude for permission to publish this paper.

REFERENCES

1 Palmer, A.C. and Baldry, J.A.S. Lateral buckling of axially-compressed pipelines. Journal of Petroleum Technology, 26, 1284-1284 (1974).

2 Schaminee, P.E.L., Zorn, N.F., Schotman, G.J.M. Soil response for pipeline upheaval buckling analysis: full-scale laboratory tests and modelling. Proceedings, Twenty-second Annual Offshore Technology Conference, Houston (1990).

3 Klever, F.J., van Helvoirt, L.C., and Sluyterman, A.C. A dedicated finite-element model for analysing upheaval buckling response of submarine pipelines. Proceedings, Twenty-second Annual Offshore Technology Conference, Houston (1990).

4 Locke, R.B. and Sheen, R. The Tern and Eider pipelines. Proceedings, 1989 European Seminar on Offshore Pipeline Technology, Amsterdam (1989).

5 Craig, I.G. and Nash, N.W. Upheaval buckling: a practical solution using hot water flushing technique. Proceedings, Twenty-second Annual Offshore Technology Conference, Houston (1990).

6 Palmer, A.C. Are bundles expensive ? Proceedings, Pipelines for Marginal Fields Conference, Aberdeen (1988).

7 Duxbury, P.G. and Hobbs, R.E. Helically-laid pipeline bundles. Proceedings, Eighth International Conference on Offshore Mechanics and Arctic Engineering, 5, 9-16 (1989).

8 Bournazel, C. Flambage vertical des conduites ensouillees. Revue de l'Institut Francais du Petrole, 55, 212-230 (1982).

9 Putot, C.J.M. Localised buckling of buried flexible pipelines. Proceedings, Twenty-first Annual Offshore Technology Conference, Houston, OTC6155 (1989).

10 Putot, C.J.M. and Rigaud, J. Upheaval buckling of buried pressurised pipelines. Proceedings, European Seminar on Offshore Pipeline Technology, Paris (1990).

11 Ellinas, C.P., Supple, W.J. and Vastenholt, H. Prevention of upheaval buckling of hot submarine pipelines by means of intermittent rock-dumping. Proceedings, Twenty-second Annual Offshore Technology Conference, Houston (1990).

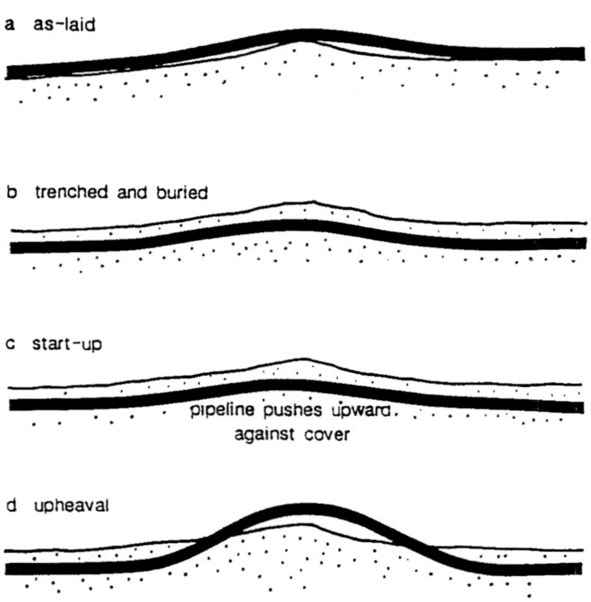

Figure 1 Sequence of laying, trenching and upheaval

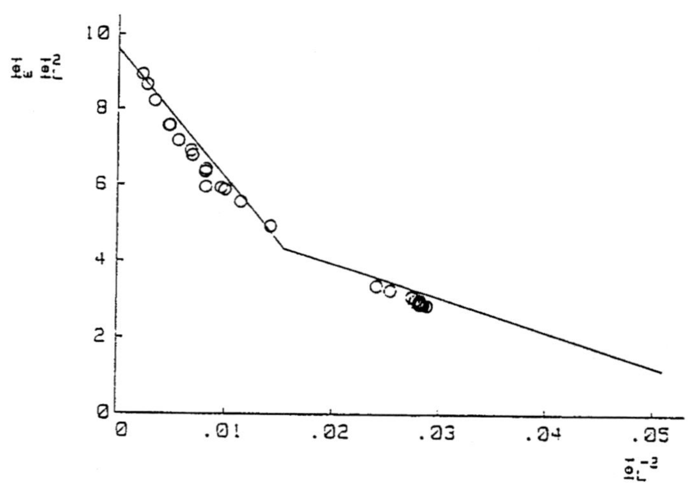

Figure 2 Numerical correlation of UPBUCK results

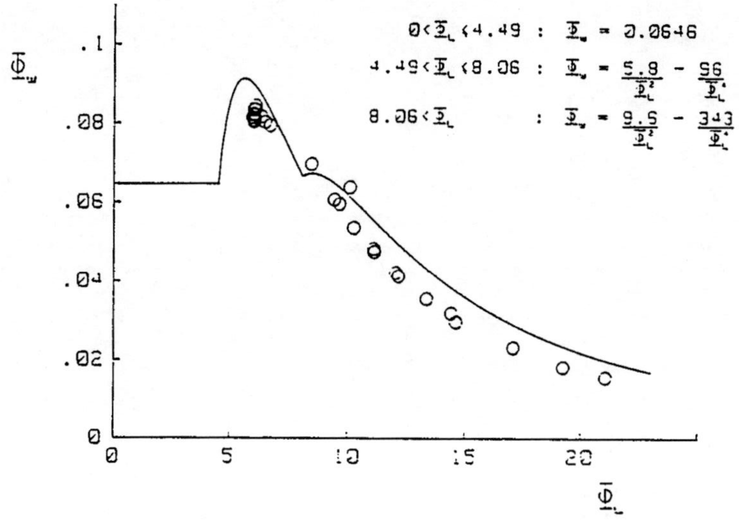

Figure 3 Universal design curve

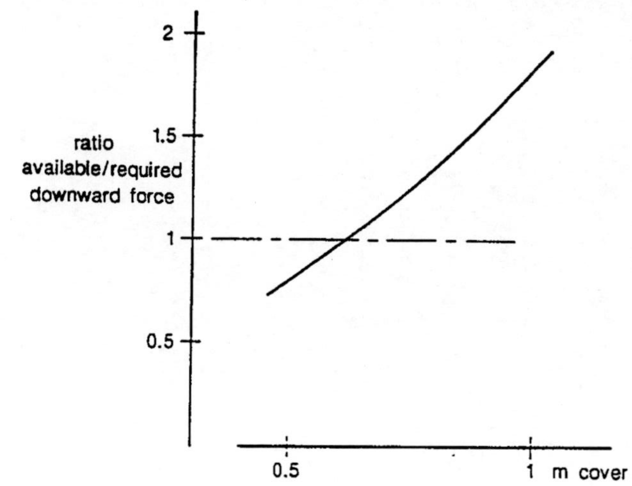

Figure 4 Design example 1: simplified method

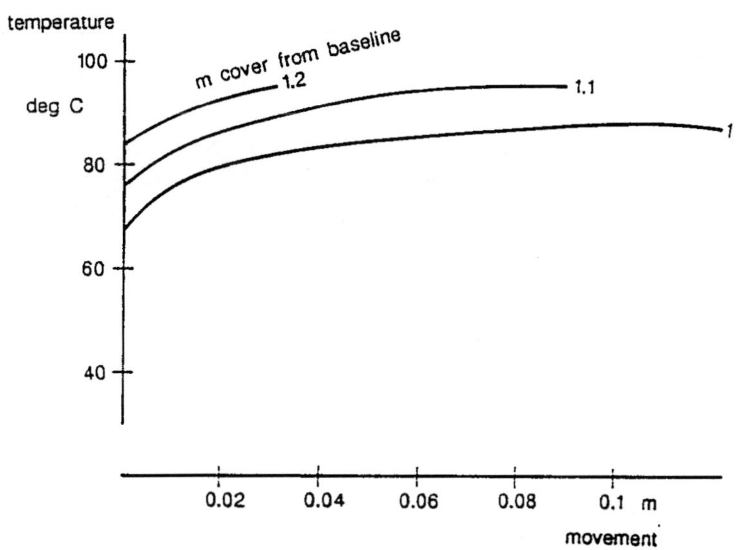

Figure 5 Design example 1 Results of level 1 UPBUCK (elastic)

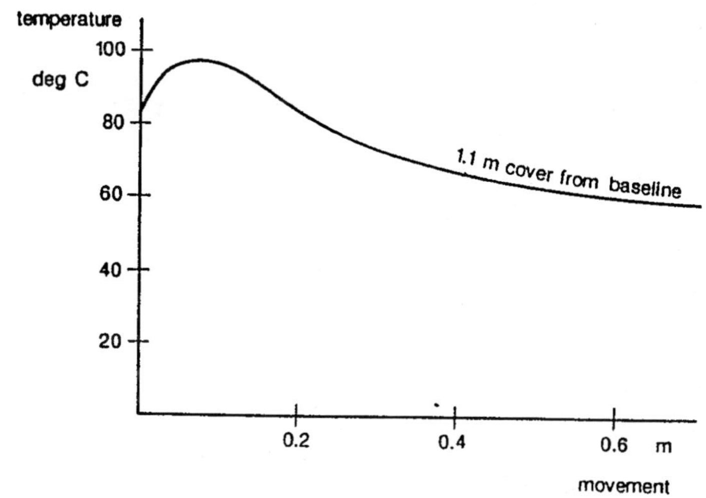

Figure 6 Design example Results of level 2 UPBUCK (elastic-plastic)

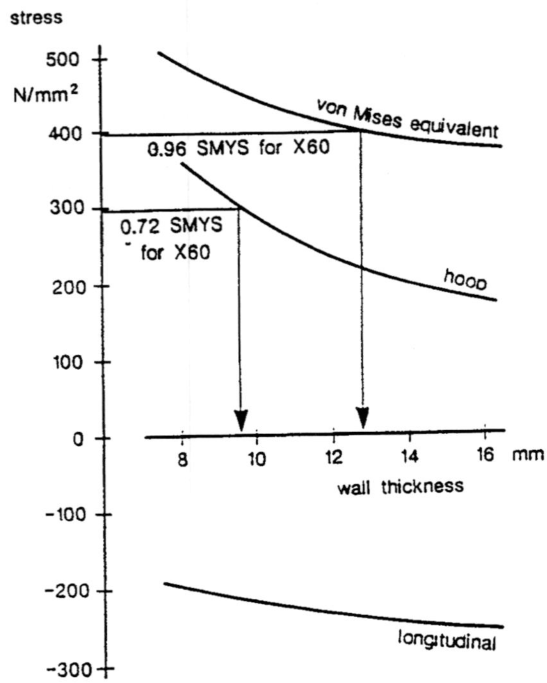

Figure 7 Example of effect of strain-based design

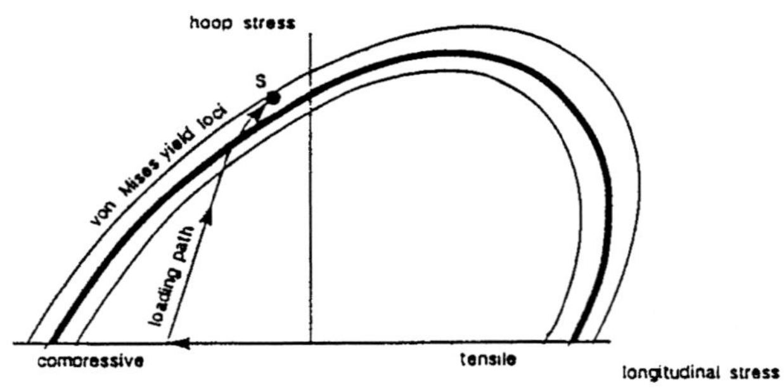

a stress path during loading beyond yield
assumes: no bending, no residual tension, temperature increase applied first, pressure increase second

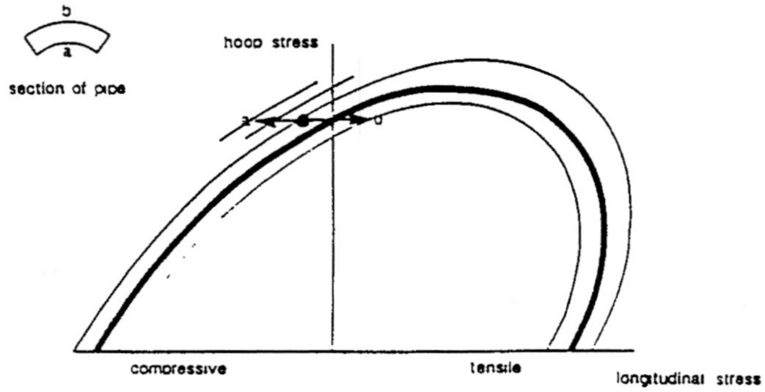

b incremental stress changes induced by bending from initial state represented by point S

Figure 8 Implications of strain-based design for stress and yield in pipeline

OTC 6831

A Preliminary Case for the Existence of Hurricane Alleys in the Gulf of Mexico

C.K. Cooper, Chevron Oil Field Research Co.

This paper was presented at the 24th Annual OTC in Houston, Texas, May 4–7, 1992.

ABSTRACT

A review of hurricane intensity in the Gulf of Mexico suggests there are two alleys where hurricanes are roughly 20% more intense than elsewhere. This paper develops a physical basis for the alleys using previous modeling and observations, and a statistical correlation of historical hurricane intensity and ocean temperatures. While the evidence for alleys is persuasive, final acceptance will depend on results of further research.

INTRODUCTION

Historically, designers and engineers have generally assumed that hurricane paths are random. This is based largely on the work of Haring and Heideman[1] and Ward et al.[2] who analyzed the ODGP wind and wave hindcast of 26 historical hurricanes in the Gulf of Mexico. While they found significant variations between sites, analysis showed that the differences were not statistically significant. In other words, the hindcast data suggested one region of the Gulf was as likely to experience a severe hurricane as another.

The recently completed GUMSHOE (GUlf of Mexico Storm Hindcast of Oceanographic Extremes) study enhanced ODGP by including more storms and shallow-water wave physics. As in ODGP, substantial spatial differences appear in GUMSHOE with three regions standing out: a mild region off central and eastern Texas, and two severe regions off Mississippi and western Texas (denoted by **L** for low, $\mathbf{H_e}$ for eastern high, and $\mathbf{H_w}$ for western high, respectively). The regions are clearly evident in Figure 1 which shows contours of the central pressure deficit (peripheral minus central pressures), a good measure of hurricane intensity. Two dark arrows superimposed on the map denote "alleys".

The pressure differences between the severe and the mild regions correspond to a 3-4 m difference in maximum wave height using the parametric model of Cooper[3]. Existing RP2A criteria suggest a spatially uniform wave height. If designers had confidence that the differences in regional severity were due to a real physical phenomena and were not statistical aberrations, they could improve the reliability of their designs. Haring and Heideman[1] and Ward et al.[2] do not provide the necessary confidence - they suggest the regional differences are statistically insignificant. However a recent analysis of the GUMSHOE statistics by Chouinard[4] suggests there is a statistically significant variation over length scales of 100 km.

A close look at the physics of hurricanes and the climatology of the Gulf waters seems to provide a physical explanation for the alleys in Figure 1. The explanation: the two alleys reflect the influence of subsurface warm water on hurricane intensity. Figure

References and illustrations at end of paper.

2 gives some support for this hypothesis. It shows the mean water temperature at 60 m depth in the Gulf of Mexico during hurricane season. A comparison of Figures 1 and 2 shows the two alleys correspond closely to the two warm water regions. The two regions in Figure 2 originate from the Loop Current and its eddies which bring warm Caribbean water into the Gulf as summarized by Elliot[5].

Hurricane experts have known for several decades that hurricane intensity is closely related to sea **surface** temperature (SST), e.g., Emanuel[6]. This is because a hurricane is a Carnot engine fueled by the temperature difference between the water surface and the air. The warmer the water the more intense the hurricane will be, all else being equal.

Unfortunately, the simple Carnot concept is not enough to explain the situation in the Gulf. The problem is the Gulf **surface** waters are nearly homogeneous in the summer. Consequently any attempt to link SST to hurricane intensity is doomed. The key to the riddle is in realizing that severe hurricanes stir up the ocean so much that they can tap the warm subsurface water shown in Figure 2. If a severe hurricane moves along one of the two alleys in Figure 1 it stirs up the water. But since there is substantial warm subsurface water, the surface does not cool much, and the Carnot engine (hurricane) is unaffected. Conversely, if the same hurricane happens to move over the Texas shelf where the subsurface water is cool, the hurricane stirs up the cooler underlying water. This cools the surface water and weakens the hurricane.

The above discussion provides a brief outline of the basis for hurricane alleys. The remainder of this paper will explore the evidence in more detail. It should be emphasized that these results are preliminary. While some of the arguments are strong, there remain questions which must be answered before spatially-variable design criteria can be used. A major goal of this paper is to outline the basis for hurricane alleys, but another equally important goal is to identify remaining questions.

The next section establishes a theoretical basis for hurricane alleys by reviewing the relevant hurricane physics established by prior work. This section is followed by one that provides evidence for the existence of warm subsurface water in the Gulf and shows how this water is correlated to hurricane intensity. The final section pulls the pieces together, summarizes the theory, examines the behavior of some historical storms in light of the theory, and identifies some lingering questions.

THEORETICAL BASIS

Ocean Response

The indigenous Gulf of Mexico is typically stratified with a 30-50 m layer of warm, constant-temperature water overlying cooler layers as depicted on the left side of Figure 3. During a hurricane, the wind stress at the surface increases, and the mixed-layer velocity increases proportionately. The underlying water remains relatively quiescent thanks to the decoupling effect of the sharp density gradient at the base of the mixed layer. Eventually, the mixed-layer velocity increases to the point that an instability occurs, and the mixed-layer entrains the cooler underlying water. The result: the mixed layer appears to deepen and cool by several degrees as depicted in the right side of Figure 3. Price[7] and Sanford et al.[8] show a drop in mixed-layer temperature of 1°-2° C can occur by the time the eye of the hurricane passes over a site. Note that the mixed-layer temperature is in practice the same as the sea surface temperature (SST) referred to frequently in the next few sections.

The cooling of the mixed layer depends strongly on several factors. All else being equal, the mixed-layer temperature will become cooler as 1) the wind stress is increased, 2) the initial mixed-layer depth is decreased, and 3) the duration of the wind stress is increased (e.g., slow storm translation speed). Figure 4 demonstrates this sensitivity for indigenous Gulf of Mexico deep water characterized by a 30-m deep mixed layer. The figure shows three temperature profiles. The "starting" curve represents pre-storm conditions. The other two are taken at the time of peak winds at a site located 1 **R** (radius to maximum winds) to the right of the storm track.

Both profiles in Figure 4 show some deepening of the mixed layer due to the entrainment process described earlier. The profile for an extreme hurricane (Camille) shows considerable deepening, and cooling of nearly 1°C of the mixed layer. The profile for a more moderate hurricane (1/2 Camille shear stress) shows much less deepening, and cooling of less than 0.2°C. A slower moving extreme hurricane like Carla would cool the mixed layer by about 2°C. It is also important to note that if the mixed layer was initially 100-150 m instead of 30 m, the change in SST would be negligible for both the severe and moderate hurricanes. The significance of these numbers will become apparent shortly. The model uses a level 2.5 turbulence closure scheme (Mellor and Yamada[9]).

In summary, observations and models show that during severe hurricanes in the Gulf of Mexico, the intense mixing of the surface layer can decrease the SST on the order of 1°C within the order of 100 km of the storm center for indigenous Gulf of Mexico water. A more moderate hurricane will decrease the SST by an order of magnitude less. Finally, the models indicate that if the initial mixed-layer depth is on the order of 100 m then the SST change near the storm center is negligible regardless of the severity of the hurricane.

Atmospheric Response

As described in Emanuel[6], a hurricane can be conceptualized as a Carnot engine with the essential elements depicted in Figure 5. In the hurricane, air spirals from *a* toward the storm center at point *c* acquiring entropy from the ocean surface of temperature T_s. The air then ascends adiabatically from point *c* flowing outward near the storm top to some large radius at point *o*. At this point, the excess entropy is radiated to the lower stratosphere at temperature T_o. The mechanical energy developed by the Carnot engine appears in the form of large wind speeds that are balanced by frictional stress at the air/sea interface.

Observations support the dependence of severity on SST (e.g., Brand[10], Fisher[11], Miller[12], and Perloth[13]). Of particular relevance to this paper is the work of Perloth[14]. He analyzed the development of hurricanes in the North Atlantic over a 65-year period and concluded that 87% of the cyclones that reached hurricane strength did so only when the difference between the sea surface temperature and water temperature at 60 m exceeded 2°C. A logical interpretation of this result is that when the mixed layer was shallow, say 30 m, the cyclone cooled the mixed layer dropping SST which in turn weakened the storm.

Coupled models also support the temperature/severity correlation suggested in Emanuel's simple Carnot model. Coupled models include both ocean and atmosphere components that account for the feedback between the two. Early coupled models, e.g., Sundquist[15], neglected storm translation and showed the peak winds dropped by 50% due to a 2°C decrease in SST. In contrast, the model of Chang and Anthes[16] suggested only a few percent drop after 24 hours. Sutyrin and Khain[17] and more recently Gallacher et al.[18] have pointed out deficiencies in the Chang and Anthes work including the lack of spatial resolution and the short integration period.

One other important insight provided by the coupled models of Sutyrin and Khain[17] and more recently Rotunno and Emanuel[19] is that the effect of SST on severity is limited to the region within the order of 100 km of the eye. It is interesting to note that Chouinard's[4] statistical analysis of the GUMSHOE results suggests a spatial coherence scale of order 100 km.

The early coupled models assumed the hurricane was stationary. This is a limiting case in which the atmosphere will be most affected by the ocean. This is because a stationary storm will cause larger changes in SST than a moving storm. Sutyrin and Khain[17,20] and Khain and Ginis[21] have developed coupled models with a translating hurricane. Figure 6, from Sutyrin and Khain[17], summarizes some of the more interesting results. It shows the time variation of central pressure for three cases: 1) V_c=7 m/s, h_m=30 m, 2) V_c=5 m/s, h_m=50 m, 3) V_c=10 m/s, h_m=50 m; where V_c is the translation velocity of the storm center and h_m is the initial mixed-layer depth. In the simulations the atmospheric model was initially started and run to equilibrium with the ocean state fixed. At time zero, the ocean was allowed to respond to the atmospheric forcing.

There are a number of interesting features in Figure 6. Case 1 with the slowest translation speed and shallowest initial mixed layer shows the most response. Other features of the response are also consistent with previous results. For instance when the storm speeds up or the initial mixed layer is increased, the pressure decrease is less noticeable.

For Case 1 in Figure 6, the total drop in maximum surface stress is 12%. While this is small, it should be pointed out that the storm is mild by Gulf of Mexico standards. One would expect more effect for stronger storms. The figure also shows that the effect of the ocean on storm severity occurs primarily during the second and third days.

One final interesting point. Previous work suggests pre-storm variations in SST can affect a hurricane's path. Brand[10] and Tisdale and Clapp[22] report observations while Khain and Ginis[21] and Chang and Madala[23] provide evidence from coupled models.

In summary, hurricane severity is closely related to SST. To first order, a hurricane is a Carnot engine that is fueled by the temperature difference between the atmosphere and ocean. This is supported by observations and models. Other important findings: 1) the SST more than the order of 100 km from the eye has little effect on storm severity, 2) the ocean response has less impact on storm severity as the initial mixed-layer or storm translation speed increases, and 3) hurricane severity may lag changes in SST by more than a day.

OBSERVATIONAL BASIS

Contours of mean subsurface sea temperature reveal two regions of warm water as shown in Figure 2 (labeled $\mathbf{H_e}$ and $\mathbf{H_w}$), and a region of relatively cool water (labeled **L**). This figure represents the average water temperature at 60 m depth during September (peak hurricane season). The data come from 51,510 temperature observations taken from 1942-1968 and compiled by Robinson[24]. The general pattern changes little during August and October, the other months in the hurricane season.

The $\mathbf{H_e}$ region reflects the position of the Loop Current, a large permanent current in the eastern Gulf. Elliot[5] describes the Loop in some detail. It consists of warm, high-salinity Caribbean water that enters the Gulf through the Straits of Yucatan, "loops" northward into the eastern Gulf then exits through the Straits of Florida where it becomes known as the Gulf Stream. On a quasi-annual cycle, the Loop penetrates so far north that it becomes unstable and a large (300 km diameter) eddy is shed. The eddy migrates westward at about 3 km/day, and eventually hits the western Gulf shelf where it slowly degenerates over the coming year.

The source of the $\mathbf{H_w}$ region is not as obvious. As shown by Vukovich and Crissman[25], Loop Current eddies have a preferred path through this region, but they should affect it only a few weeks in the year. It is surprising that $\mathbf{H_w}$ is not located about 2° farther to the northwest where data and models (e.g., Lewis et al.[26], Brooks and Elbe[27], Hurlburt and Thompson[28]) consistently show eddies become trapped on the western shelf and persist for many months. One possible explanation for the position of $\mathbf{H_w}$ is lack of data - Robinson[24] does not indicate the data density. Given the lack of commercial interests in the western Gulf, it is reasonable to expect that the data is sparse, and consequently inaccurate. This is supported by a review of the NODC data archive for 1942-66 - there are only 10 months when data were available west of 95°W, and only 3 comprehensive cruises.

It is of more than passing interest to note that the $\mathbf{H_e}$ and **L** regions in Figure 2 lie nearly on top of their counterparts in Figure 1. There also appears to be some correlation between the $\mathbf{H_w}$ regions although the $\mathbf{H_w}$ in Figure 2 is a bit farther to the southeast than $\mathbf{H_w}$ in Figure 1.

Figure 7 quantifies the visual correlation by plotting the subsurface temperature versus the central pressure. The good correlation coefficient of 0.6 confirms the earlier impression of Figures 1 and 2. The correlations for other measures of storm severity and subsurface temperature are equally strong. The fit in the figure is especially good when one considers that it is based on average conditions, not the actual conditions at the time of the storm. Each point in the figure is based on a spatial average over a circle of 150 km in diameter. This averaging eliminates high frequency noise without eliminating the spatial scales of interest. The figure does not include the regions 1) east of 85°W or south of 25°N primarily because this is not of interest to the oil industry; 2) sites on the shelf because the hurricane could have been influenced by land; and 3) data west of 95°W because the climatological temperature data are too sparse.

In summary, hurricanes appear to be more intense in a western and eastern region, and less severe in between. The eastern region corresponds closely to a region of warm subsurface water originating from the Loop Current. The less severe region corresponds to a region of cool, indigenous Gulf water.

DISCUSSION

Putting the Pieces Together

A hurricane is a Carnot engine fueled by the thermodynamic disequilibrium between air and ocean. The hurricane's intensity is proportional to the sea surface temperature (SST) - the intensity will drop significantly if the SST drops by the order of 1°C according to simple but realistic models. On the ocean side, measurements and models show that the SST will drop by this same order if a severe storm travels over indigenous Gulf waters. It follows that a severe storm will tend to weaken in this case.

Conversely, the models and measurements show the SST will be largely unaffected if a severe storm travels over the warmer Loop Current waters. It follows that a severe storm can be maintained or perhaps even intensify in this case. Finally, a less severe storm will be largely unaffected by the type of water.

These expectations are supported by a preliminary review of historical hurricane severity and climatological temperature data. An analysis shows the severity of the storms is well correlated to subsurface temperature.

A note of caution - the hurricane alley theory does not say a severe hurricane cannot occur outside the alleys. Recall that Figure 2 is based on climatological averages and major variances can occur. For instance, a Loop Current eddy could migrate through the region at the same time as a severe hurricane. Since the eddy contains warm water, the theory states a severe hurricane could persist. But since the likelihood of an eddy being present is low, the theory suggests that the severity of hurricanes will be less on average.

Individual Hurricanes

The discussion thus far has focused on average conditions. An obvious question remains: Are there severe individual storms that tracked over regions of cool subsurface water without weakening? There would be strong reason to doubt the existence of alleys if the historical record contains this class of storm. Unfortunately getting a black and white answer to the question is not easy for several reasons. First, it requires detailed measurements of ocean temperature at or near the time of the individual storm and these may not be available. Second, previous modeling has not adequately defined "severe" or "slow-moving". Despite these restrictions, a brief look is in order. For now, we focus on the 15 most severe storms passing within 500 km of Galveston. The tracks are based on the NOAA archive covering 1900-1990.

Figure 8 shows the tracks of the five most severe storms. The numbers near the coast denote the central pressure at that position in mb. All the storms with the exception of Carla track close to the two alleys depicted in Figure 1. A review of the NODC ocean temperature suggests Carla passed over an eddy near the southwestern Texas shelf. An eddy would have provided the necessary deep hot water needed to sustain such an intense and large storm like Carla. The NODC archive also shows eddies were present in the western Gulf during Allen and Beulah. Finally, the archive suggests the Loop Current extended well to the north during Camille. No data are available for the 1915 storm.

The situation is less clear as the severity of the storm decreases. Figures 9 and 10 show the storms ranked 6-15 in severity. Unfortunately there are not any good ocean temperature data for these storms except for Gilbert which intensified over a warm eddy just before going ashore. Despite the lack of temperature data a review of the nine other storms using climatological temperatures is worthwhile because it raises some interesting questions.

Five of the storms followed the alleys where warm subsurface water is normally found. Two others, Celia and the 1919 storm, tracked near the Texas shelf break where eddies are sometimes found. Hence 7 of the 10 storms seem to follow a trend consistent with the alley hypothesis. However, Audrey, and the Galveston and 1932 storms seem to challenge the hypothesis. They tracked perpendicular to the shelf over regions where the Loop Current and eddies are never found, and they remained intense right up to the coast. There are potential explanations.

1. The storms were not severe enough or moved too fast to be substantially affected by the oceanic response. For instance Audrey was moving at 14 kt when it tracked over the shelf. Recall that the selection of the 15 severest storms is somewhat arbitrary given our poor quantitative definition of "severe" and "fast-moving".

2. The intensities of the storms were overestimated. Two of the storms occurred before 1952 when aircraft reconnaissance became routine. Prior to reconnaissance, intensity estimates were based on scattered ships of opportunity and land stations, and are much less reliable than after 1952.

3. The shelf waters were unusually warm. Recall that Figure 2 is based on climatological means and there will be some year-to-year variability. For instance Temple et al.[29] show that in 1963 temperatures at 60 m reached 26°C, about 4°C warmer than usual.

In summary a look at the 5 most severe storms in the Gulf supports the existence of alleys. The case for the less severe storms is more ambiguous. Temperature data is only available for Gilbert and it supports the alleys. Of the remaining 9 storms only three appear to challenge the alley hypothesis, and consequently deserve a closer look.

Future Research

The case for hurricane alleys in the Gulf is compelling but remains somewhat circumstantial. Several questions remain, and must be answered before designers can take advantage of alleys. These questions include:

1. How could the 1900, 1932, and 1957 (Audrey) storms propagate between alleys? It has been suggested that these storms were too weak or moved too fast to be affected by the oceanic response. These two explanations can best be explored hindcasting the storms with a proven coupled model.

2. Is there a 1-3 day lag between changes in SST and severity as suggested by the models of

Sutyrin and Khain[17,20]? If so this would mean that a hurricane could travel on the order of 300 km before its severity is affected by the ocean response. In a basin as small as the Gulf this would mean the ocean response has a marginal impact on severity.

3. Is the temperature pattern shown in Figure 2 accurate? A close look at the bases for the figure in the western Gulf revealed little actual data. The question is important because if the pattern is not fairly stable from year to year then the hurricane alleys won't be either.

4. How reliable are the estimates of severity for the pre-reconnaissance storms? If they were overestimated this would reduce the criteria by roughly 20% since these storms dominate the extremal analysis. The basis for the severity of these early storms seems to be poorly documented.

The first and second questions are the focus of a recently initiated one-year study by MIT and Chevron. The study is funded by eight oil companies: Amoco, Chevron, Conoco, Exxon, Marathon, Mobil, Shell, and Texaco. Research on the other two questions is expected to begin this year.

REFERENCES

1. R. C. Haring and J. C. Heideman: Gulf of Mexico Rare Wave Return Periods, Proc. Offshore Tech. Conf., 3230, 1537-1550, 1978.

2. E. G. Ward, L. E. Borgman, and V. J. Cardone: Statistics of Hurricane Waves in the Gulf of Mexico, Proc Offshore Tech. Conf., 3229, 1523-1536, 1978.

3. C. K. Cooper: Parametric Models of Hurricane-Generated Winds, Waves, and Currents in Deep Water, Proc. Offshore Tech. Conf., 5738, 475-484, 1988.

4 L. E. Chouinard: A Statistical Method for Regional Design Wave Heights in the Gulf of Mexico, Proc. Offshore Tech. Conf., 6832, 1991.

5. B. A. Elliot: Anticyclonic Rings in the Gulf of Mexico, J. Phys. Oceanogr., 12, 1292-1309, 1982.

6. K. A. Emanuel: Toward a General Theory of Hurricanes, Am. Scient., 76, 371-379, 1988.

7. J. F. Price: Upper Ocean Response to a Hurricane, J. Phys. Oceanogr.,11 , 153-175, 1981.

8. T. B. Sanford et al.: Ocean Response to a Hurricane, I: Observations, J. Phys. Oceanogr., 17, 2064-2083, 1987.

9. G. L. Mellor and T. Yamada: A Hierarchy of Turbulence Closure Models for Planetary Boundary Layers, J. Atmosph. Sciences, 31, 1791-1806, 1974.

10. S. Brand: The effects on a Tropical Cyclone of Cooler Surface Waters Due to Upwelling and Mixing Produced by a Prior Tropical Cyclone, J. Appl. Meteo., 10, 865-874, 1971.

11. L. Fisher: Hurricanes and the Sea-Surface Temperature Field, J. Meteorology, 15, 329-333, 1958.

12. B. I. Miller: On the Maximum Intensity of Hurricanes, J. Meteorology, 15, 184-195, 1958.

13. I. Perloth: Hurricane Behavior as Related to Oceanographic Environmental Conditions, Tellus, 19, 258-268, 1967.

14. I. Perloth: Effects of Oceanographic Media on Equatorial Atlantic Hurricanes, Tellus, 21, 230-244, 1969.

15. H. Sundquist: Mean Tropical Storm Behavior in Experiments Related to Modification Attempts, Tellus, 24, 6-12, 1972.

16. S. W. Chang and R. A. Anthes: The Mutual Response of the Tropical Cyclone and the Ocean, J. Phys. Oceanogr., 9, 128-135, 1979.

17. G. G. Sutyrin and A. P. Khain: Effect of the Ocean-Atmosphere Interaction on the Intensity of a Moving Tropical Cyclone, Investiya, Atmos. and Oceanic Physics, 20, 697-702, 1984.

18. P. C. Gallacher, R. Rotunno, and K. A. Emanuel: Tropical Cyclogenesis in a Coupled Ocean-Atmosphere Model, 18th Conf. on Hurricanes and Tropical Meteor., San Diego, Amer. Met. Soc., 1989.

19. R. Rotunno and K. A. Emanuel: An Air-Sea Interaction Theory for Tropical Cyclones. Part II: Evolutionary Study Using a Nonhydrostatic Axisymmetric Numerical Model, J. Atmosph. Sci., 44, 524-561, 1987.

20. G. G. Sutyrin and A. P. Khain: Interaction of the Ocean and the Atmosphere in the Region of a Moving Tropical Cyclone, Dokl. AN SSSR, 249, 467-470, 1979.

21. A. Khain and I. Ginis: The Mutual Response of a Moving Tropical Cyclone and the Ocean, manuscript submitted for publication, 1991.

22. C. F. Tisdale and P. F. Clapp: Origin and Paths of Hurricanes and Tropical Storms Related to Certain Physical Parameters at the Air-Sea Interface, J. Appl. Meteorology, 2, 358-367, 1963.

23. S. W. Chang and R. V. Madala: Numerical Simulation of the Influence of Sea Surface Temperature on Translating Tropical Cyclones, J. Atmos. Sci., 37, 2617-2630, 1980.

24. M. K. Robinson: Atlas of Monthly Mean Sea Surface and Subsurface Temperature and Depth of the Top of the Thermocline, Gulf of Mexico and Caribbean Sea, Scripps Institution of Oceanography, March, 1973.

25. F. M. Vukovich and B. W. Crissman: Aspects of Warm Rings in the Gulf of Mexico, J. Geophys. Res., 91, 2645-2660, 1986.

26. J. K. Lewis et al.: Evolution of a Warm-Core Ring in the Gulf of Mexico: Lagrangian Observations, J. Geophys. Res., 94, 8163-8178, 1989.

27. D. A. Brooks and M. C. Elbe: Moored Array Observations in the Western Gulf of Mexico, Rep. 82-12-T, Tex. A&M Univ., College Station, Oct., 1982.

28. H. H. Hurlburt and J. D. Thompson: A Numerical Study of Loop Current Intrusions and Eddy Shedding, J. Phys. Oceanogr., 10, 1611-1651, 1980.

29. R. F. Temple, D. L. Harrington, and J. A. Martin: Monthly Temperature and Salinity Measurements of Continental Shelf Waters of the Northwestern Gulf of Mexico, ·1963-65, NOAA Tech. Rep. NMFS SSRF-707, 28 pp, 1977.

ACKNOWLEDGEMENTS

J. Lewis of SAIC is acknowledged for his stimulating suggestion during the 1988 MMS Gulf of Mexico Studies meeting that hurricanes could be influenced by the Loop Current. L. Chouinard of Chevron is acknowledged for his efforts to show the statistical relationship between hurricane severity and water temperature. J. Olesberg assisted in preparing figures. K. Schaudt provided plotted summaries of the NODC temperature. Thanks is also given to K. Emanuel of MIT, V. Cardone of Oceanweather, and C. Petrauskas of Chevron for helpful discussions.

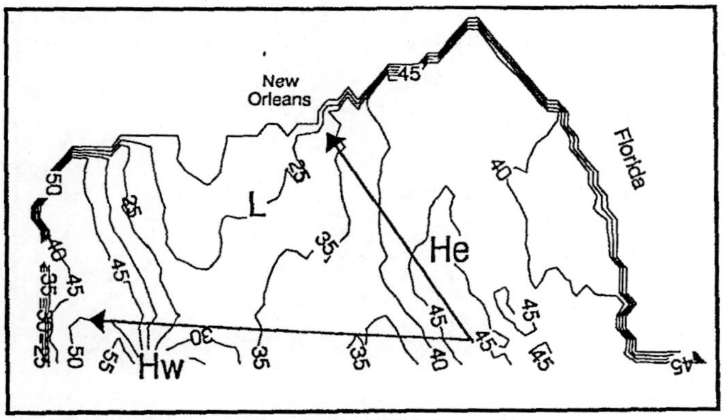

Figure 1: Spatial variation of central pressure deficit (mb) in the Gulf of Mexico for hurricanes from 1900-1989.

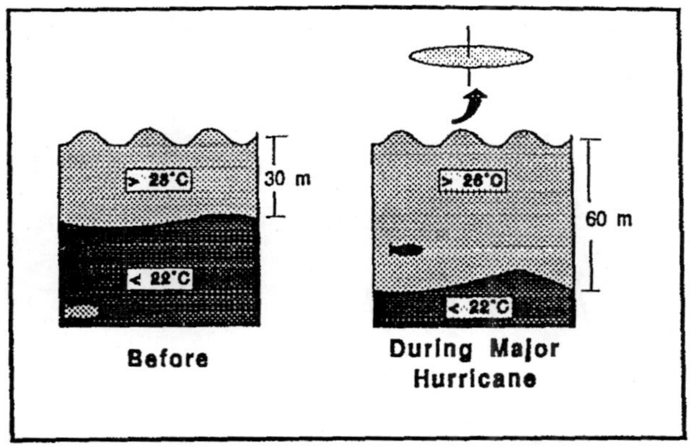

Figure 3: Typical Gulf of Mexico water column in deep water before and after a major hurricane.

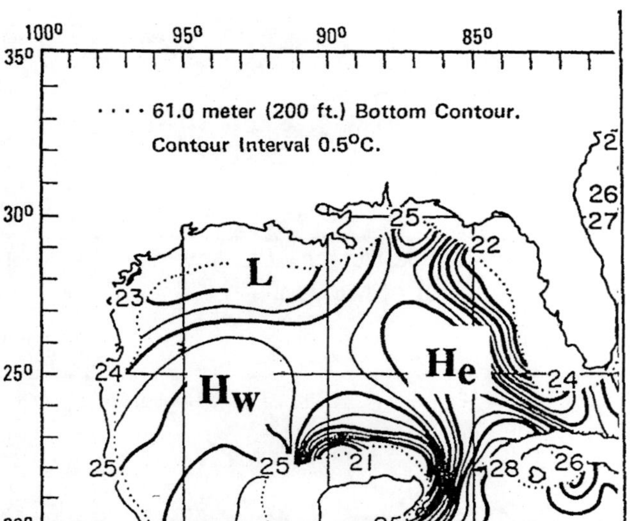

Figure 2: Climatological temperature at 60 m during September in the Gulf of Mexico. Figure is from Robinson[24].

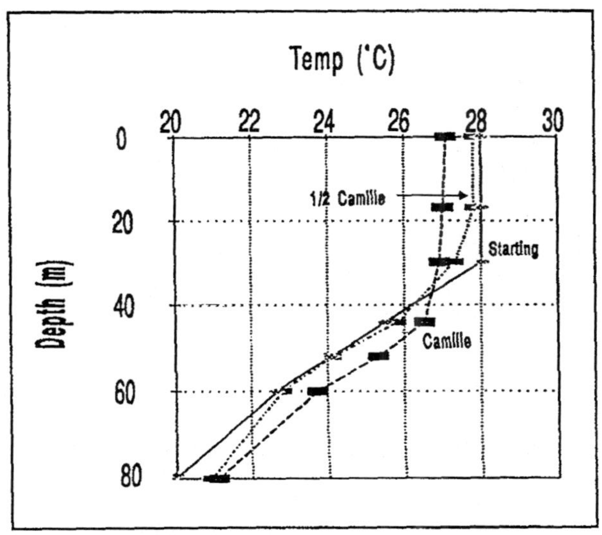

Figure 4: Temperature profiles prior to and during severe and moderate hurricane.

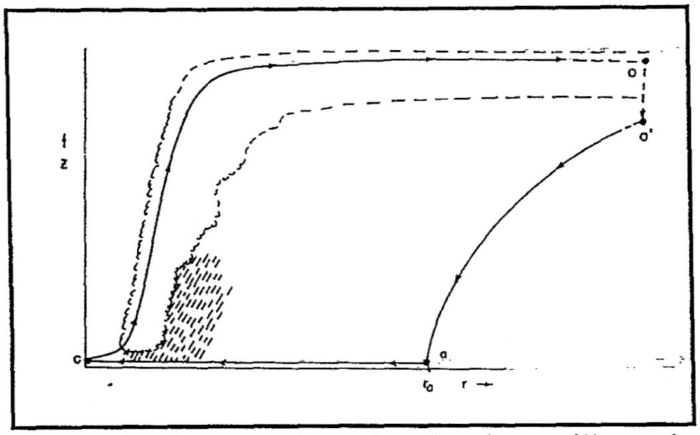

Figure 5: Cross-section of a hurricane illustrating the basic elements of the Carnot cycle (from Emanuel[6]).

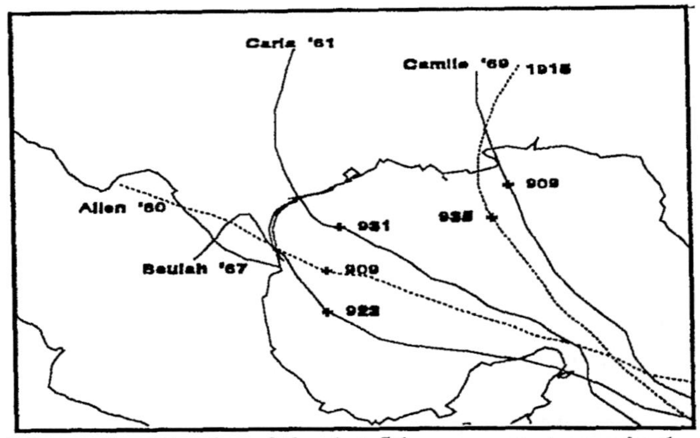

Figure 8: Tracks of the 1st-5th severest storms in the Gulf of Mexico.

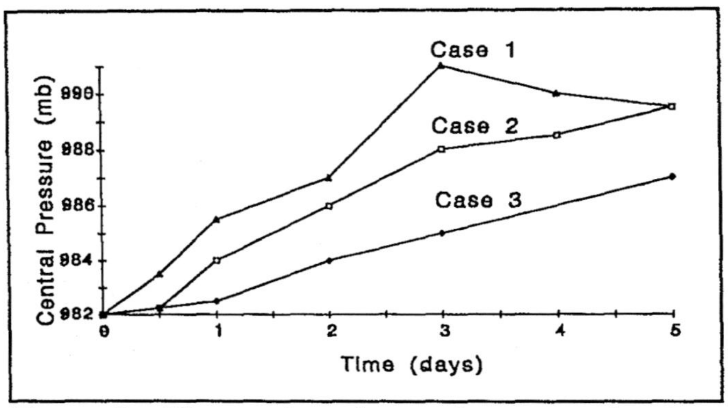

Figure 6: Time series of central pressure for three runs of the numerical model of Sutyrin and Khain[30].

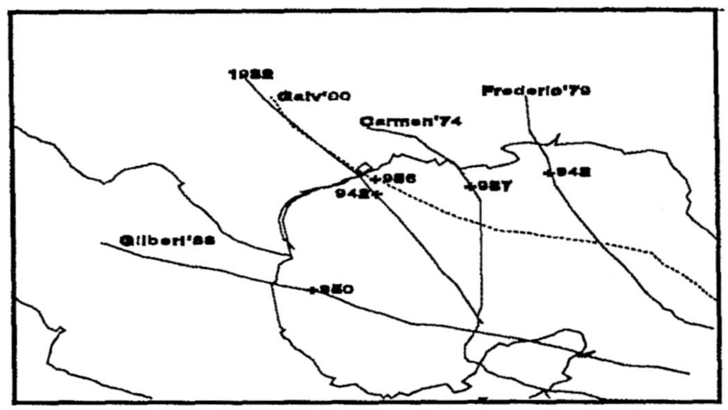

Figure 9: Tracks of 6th-10th severest storms in the Gulf of Mexico.

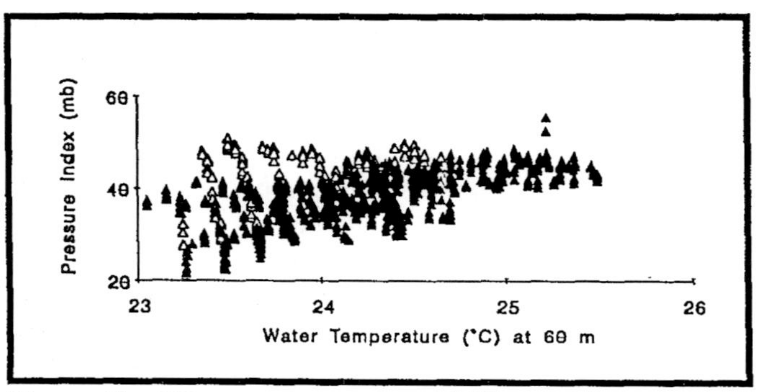

Figure 7: Water temperature (°C) at 60 m versus hurricane pressure index (central pressure minus peripheral pressure in mb).

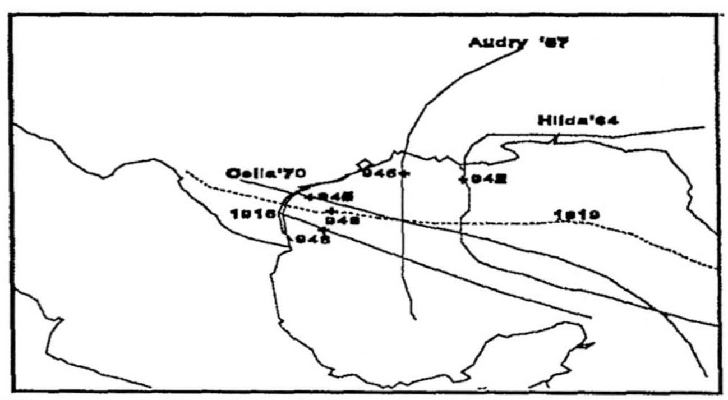

Figure 10: Tracks of the 11-15th severest storms in the Gulf of Mexico.

OTC 6965

Appraisal of Lightweight Moorings for Deep Water

C.R. Chaplin, Reading U., and C.J.M. Del Vecchio, Petrobrás S.A.

This paper was presented at the 24th Annual OTC in Houston, Texas, May 4-7, 1992.

Abstract

This paper outlines the potential advantages of using lightweight fiber ropes for mooring floating production platforms. The influence of tether properties on mooring system response is discussed. It is shown that a comparison of alternative ropes in terms of their mechanical properties is inadequate; relative merit can only be assessed through comparison of full systems that have been designed to minimum cost, and to common performance specifications. For catenary systems, it is found that a lightweight option using polyester fiber can outperform both traditional and more exotic solutions. The significant mechanical performance characteristics governing mooring systems design have been measured for a model tether.

Introduction

Work on this topic was initiated at Reading University in 1989 with support through a joint industry project on "lightweight materials for offshore station keeping" [1]. The primary objective of that study had been to rank the relative merits of mooring components manufactured from lightweight materials using available fiber and rope technology, in relation to conventional offshore platform systems. The investigations carried out in the course of that study demonstrated that simplistic comparison of rope properties was potentially misleading. To facilitate any meaningful comparison it was found essential to perform an analysis of the complete mooring system, optimised in terms of cost and designed to comparable criteria.

References and figures at end of paper

The function of a mooring system is to restrict the lateral motions of the floating structure within specified operational constraints. Thus the overall design parameters are typically in terms of maximum permissible offsets, varying according to the operating conditions and the riser specification, with limitations on maximum tether loads and other more specific constraints on say anchor uplift.

The design loads on the overall system are generated by wind, waves and currents, all of which are complex functions of time and stochastic in nature.

Essentially any mooring system operates by developing horizontal components at the points of attachement to the floating structure. The loads in the mooring system will change in response to lateral displacements, thereby generating the restoring forces. The primary performance criterion of a mooring system is therefore its lateral stiffness – its station keeping performance. The details of the system performance are a function of the interactive frequency response of the vessel and mooring system. The loads induced in the mooring components as a result of the platform motions, in turn determine:

(i) the safety factor between component strength and the maximum loads under extreme conditions; and,

(ii) the spectrum of repeating loads in relation to fatigue performance.

Mooring concepts

For operating in deepwater there are two alternative mooring concepts which can be broadly categorised as catenary or TLP. The stiffness and frequency response of the two classes are fundamentally different, particularly in the first order wave frequency range. Because of these differences and the way in which the mechanical properties of the mooring components influence loads and therefore design, the two types of system must be considered separately.

Catenary

The catenary mooring system for semi-submersible types of floating structure can involve different configurations with variations including:

(i) conventional anchors or piled anchors;

(ii) spring buoys, clump weights or heavy grounding chains;

(iii) conventional fairleads, bending shoes, J tubes; and

(iv) a whole range of options in relation to the type and position of winches.

However the common essential characteristic is a system with long natural periods for motions in all modes.

The floating structure moves laterally in response to the environmental loads. It will then oscillate about a mean offset position. The compound motions will typically comprise two distinguishable frequency components corresponding to the fundamental, first order, higher frequency component of the waves, and the second order, low frequency, component resulting from grouping of the waves and wind. The typical response of semi-submersible structures, in all but the shallowest depths, is such that the short period wave frequency motions are, in effect, unrestrained by moorings. The characteristic amplitudes of the combined motions are such that for a semi-submersible floating production system, the well heads must be located on the sea bed, and risers must have an element of flexibility to accommodate these motions.

The design procedure for a mooring system must limit the extreme combination of the three independently calculated offsets associated with steady state (mean), low frequency and high frequency.

Procedures in API RP 2FP1 [2], set out guidelines for the design of moorings for floating production systems. Under extreme operating conditions, the maximum tensile loads should be less than 60% of the breaking load of the installed mooring line. These maximum loads are calculated by adding together three components: the mean tension, a component related to the low frequency loading, and a component related to the first order wave frequency component. For these two oscillatory components, the force of greater amplitude is added in at full value, but the lower amplitude component is added in at "significant" level (i.e. the mean of the highest third of the range). This procedure compensates for the improbability of the two peaks coinciding.

However, the more important aspect here is the distinction between the components at high and low frequencies. Ignoring the complications of spring buoys and the like, and thinking for the moment only in terms of conventional steel tethers, at the lower frequencies the line behaves in an essentially quasi-static manner: there is time for the catenary to change shape, the grounding point of the line to move, and the overall compliance (and therefore the station keeping performance) is determined by the water depth, the weight of the line and the mean tension. The stiffness of the line itself has only a limited influence on the mooring performance.

For loadings at increasingly higher frequency the situation changes. Indeed the loading is really a displacement, the motion induced in the floating structure by the waves, and the forces in the line reacting to these motions are determined by the effective line stiffness. For motions of increasing frequency, the hydrodynamic components acting on the line become increasingly significant so that forces develop lateral to the line restricting changes in catenary shape. The consequence is an increase in the effective line stiffness which tends towards the axial stiffness of the tether itself [3]. Where big waves are a frequent occurrence, this may lead to significant fatigue components.

From the design point of view the frequency response of the conventional catenary system is the wrong way round. The offset limitations tend to dictate the necessary quasi-static stiffness, while the stiffness at wave frequency determines the fatigue loading. The increase in stiffness with frequency is therefore a significant disadvantage.

This situation changes completely when lightweight materials are used. The essential feature of such materials in this context is that the tether as "packaged" is close to neutral buoyancy. The type of mooring configuration which we have considered here would comprise of a lightweight fiber tether attached at the vessel end to a more rugged steel component operating through a conventional fairlead to a conventional winch (Figure 1). At the lower end we have considered two alternative configurations: attachment via a chain to a conventional anchor, or attachment directly to a pile which can sustain any applied uplift.

So, for a neutrally buoyant tether, from the vessel to the ground chain, or pile, the line is essentially straight. Catenary shape change is only relevant to the ground chain, axial line stiffness becomes much more import, and the difference between high and low frequency stiffness is reduced. As a result, the mooring stiffness can be controlled to a significant degree through the choice of material, having regard to the working strains. And for comparable offset response at low frequencies, fatigue components due to wave frequency motions are appreciably lower.

When comparing lightweight mooring systems with conventional wire rope systems (at least those without spring buoys) there are additional advantages to be obtained due to the fact that for any given water depth the lightweight line approaches the vessel at a much flatter angle. So for any given horizontal force reacting the environmental loading, the tension that must be developed in the mooring line is appreciably lower, as also is the vertical loading imposed by the moorings on the vessel. Such vertical components can become very significant in deep water, representing the equivalent of substantial deckloading, or looked at from another point of view, a substantial cost in terms of buoyancy provision.

The benefits outlined here will be seen in the example designs analysed below. The main point at this stage though is to explain how any meaningful comparison of the relative merits of alternative materials can only be seen through a detailed analysis of the application for which they are intended; in this case deepwater mooring systems. The other point to emerge from the discussion is the role of line stiffness: for any specific installation there is probably an optimum stiffness for providing the desired sation keeping performance while minimising fatigue loadings, and exploiting the full load bearing potential of the tether. The difficulty of simultaneously matching all these requirements, and minimising cost, will also be evident in the case studies.

TLP

It would be inappropriate to ignore the tension leg concept in any discussion of moorings for deep water. The primary advantage of the TLP concept is seen in terms of limiting vertical motions to a level that permit the use of "rigid" risers with the well heads located on the platform. The requirement for high vertical restraining forces developed in parallel pre-tensioned vertical tethers results in a system with very much higher natural frequencies for modes with vertical components (heave, pitch and roll). These lower natural periods lead to the need to prevent resonance at the first order wave frequency. In the first generation of TLP designs this problem was overcome by ensuring that the natural heave period was appreciably less than typical wave period, say a maximum of 3 seconds [4]. This approach led to a stiffness requirement for the tethers, such that at higher depths the working stresses were restrained making the use of higher tensile components, such as wire rope products, unacceptable.

Changes in the TLP concept, with very much greater draft coupled with smaller water plane areas and other subtle design features, have significantly reduced the driving forces at wave frequency, so leading to some relaxation in the stiffness requirements. Nevertheless the trend is still very much towards high stiffness. In these terms the only prospective alternatives to tubular steel tethers, or wire rope constructions, would seem to be the higher stiffness composite types of structures [5], comprising bundles of pultruded carbon fiber rod, or composite tubulars. However a more radical approach would be to opt for low stiffness tethers, so avoiding the resonance problem by ensuring the natural heave periods are appreciably higher than the first order wave periods. Such a radical approach could open the way for low cost lightweight TLP tethers which because of their low stiffness would have the additional advantage of very low fatigue load amplitudes.

Analysis

The joint industry study mentioned above [1], involved a preliminary evaluation of lightweight tethers for catenary moored semi-submersible platforms. The starting point of this evaluation was a survey of tether properties relevant to mooring design, including cost. Physical and mechanical properties were obtained from a critical survey of the literature, complemented by a number of in-house tests. Cost figures were mostly obtained from manufacturers' quotations based on the supply of 8 mooring lines of 500 tonf MBL, each 1000 m long. The effect of jacketing on either properties or cost was not considered. Table 1 provides an extract of the findings of this survey.

Using a standard 8 leg anchor pattern, of the three component type shown as (b) in Figure 1, lightweight systems were compared with wire rope/chain combinations. Each configuration was optimised for cost, based on a quasi-static analysis, but with dynamic effects simulated by using tether axial stiffness to approximate wave frequency response. The outcome of this analysis showed that systems based on ropes with aligned constructions using polyester yarn could provide cost competitive solutions in water depths in the range of 500 to 2000 m. Through a combination of low weight, favourable geometry and moderate dynamic stiffness, systems of lower breaking load were able to meet the same safety factors of wire/chain moorings.

Systems based on ropes made of stronger and stiffer fibers were only able to meet the same safety factors at substantially higher costs. The analysis highlighted the importance of performing a dynamic analysis incorporating accurate figures for tether stiffness. It was also apparent that systems incorporating polyester ropes had very small wave frequency tension amplitudes when compared with wire/chain combinations.

From a consideration of possible installation and operation procedures for three component mooring lines (wire or chain, light weight rope, and ground chain), it became clear that long term strength, stiffness and creep would dominate design considerations.

The significant system failure modes to be addressed are:

(i) tensile overload;

(ii) creep rupture; and,

(iii) contact between the fairlead and the connection between the light weight tether and the upper steel section.

An experimental program was developed to assess the variation of the properties relevant to these perceived failure modes. The experimental work is discussed in the following section.

To obtain a better measure of the advantages of using polyester ropes in deep water moorings, a series of case studies was investigated based on an Aker H3.2 platform used as a Floating Production System. This system was analysed for three water depths (500, 1000 and 2000 m) in two locations (Campos Basin and West of Shetlands). The three alternative mooring systems shown in Figure 1 were considered:

(i) polyester parallel-strand rope attached to a pile;

(ii) polyester parallel-strand rope with ORQ (oil rig quality) ground chain and drag embedment anchor; and,

(iii) six-strand wire rope also with ORQ chain and anchor.

For the 1000 m West of Shetlands case, a further combination was included, consisting of standard modulus aramid parallel-strand rope with ORQ chain and anchor.

An 8 line 45° symmetric mooring pattern, operated passively, was assumed, with quartering seas and current varying linearly with water depth. The evaluation was based solely on performance under maximum design condition for intact systems as defined in API RP 2FP1 [2]. The mooring analysis procedure used involved a deterministic approach with frequency domain analysis of line tensions, following a sequence of:

(i) static analysis under mean environmental loads;

(ii) computation of wave frequency motions;

(iii) computation of low frequency loads and motions;

(iv) derivation of revised equilibrium position and corresponding line tensions incorporation the low frequency component, taking into account the worst combination of low frequency and wave frequency responses;

(v) frequency domain wave frequency analysis of line tensions; and,

(vi) combination of static, low and wave frequency tensions and offsets for evaluation against design criteria.

It should be noted that:

(i) for systems using polyester ropes, different values for static, low and wave frequency axial stiffness were used;

(ii) maximum tensions were limited to 60% of MBL and offsets to 10% of water depth; and,

(iii) through iteration designs were optimised to minimise the capital cost of tethers.

The results of this exercise are sumarised in Tables 2 and 3, for Campos basin and the West of Shetlands respectively.

Figure 3 shows the the tensions in all cases analysed. The trends shown by the results for both locations are similar in that much lower tensions are obtained on the systems using polyester ropes. This results in configurations using ropes of much lower MBL.

It should be noted that because of the higher wave loading, the differences are greater West of Shetlands where, in 1000 m, maximum tension in the polyester rope/chain system is only 61% of the tension on the wire rope/chain system. It is worth noting that West of Shetlands at 500 m water depth, there would be an advantage in using ropes with an even lower axial modulus.

An effort was made to design all the systems to use the full offset allowable (10%), but, as shown in Figure 3, it was not always possible to obtain minimum cost solutions for the polyester systems in this way, so the bonus of a low offset is obtained.

The systems with piles performed very similarly to the systems with polyester ropes and chain in all load cases except in 500 m in the West of Shetlands, due to its excessive stiffness. It should be noted that the piled systems obtained lower offsets, so without compromising cost, 6% is readily achieved in 1000 m water depth.

The capital cost of the tether components of alternative systems is shown in Figure 3. Even without considering the economies that follow from the lower vertical forces on the floating structure (roughly half with synthetic ropes), systems based on polyester rope/chain combinations are shown as cheaper than wire rope/chain combinations. Component costs are typically 80% of the conventional alternative for West of Shetlands, and 90% for Campos basin. Polyester/pile systems work out even cheaper (typically 50% of a conventional system), but the comparison between the cost of piles and anchors is not so straightforward.

For both the synthetic and steel wire rope combinations with chain, it is thought that differences in installation cost would not be significant.

The system based on an aramid rope/chain combination had the highest cost and the highest wave frequency tensions, but it can be tailored to obtain very low offsets. At approximately 60% premium on the cost tabulated, a system can be designed with maximum offset as low as 3% in 1000 m water depth.

Material and Construction Properties

Tensile strength

Strength was evaluated on model ropes of two constructions with approximately 60 kN breaking load:

(i) a torque balanced 14 x three-strand rope construction with braided cover (*Superline*® from Marlow Ropes Ltd.); and,

(ii) a torque balanced 34 strand parallel-strand rope with braided cover (Brascorda Parallel from Brascorda SA).

The core diameter of both ropes was approximately 11.5 mm with linear weights of 0.0995 kg/m and 0.107 kg/m respectively.

Testing procedure included an installation cycle at 50% of UBL and 100 cycles at 20 ± 10% of UBL, for construction bedding in. Measured ultimate breaking loads (UBL) were found to be 57.2 kN and 58.7 kN for the *Superline*® rope and for the Brascorda Parallel rope respectively. These values indicate a strength efficiency of approximately 75% on a linear weight basis for both ropes.

Axial stiffness

Relevant frequencies for stiffness measurements range from the shortest wave period (5 seconds, or 0.2 Hz) to a one week period (1.65×10^{-6} Hz) corresponding to reversals in prevailing environmental conditions.

Mean tether tension was varied from 5 to 50% of MBL and tension range from 0 to ± 15% of MBL.

For mooring systems designed with a steel upper component, an "installation" stiffness is important. Therefore measurements were also made of secant modulus between zero load and 50% of MBL held for 30 minutes (as shown in Figure 2).

Axial stiffness (EA) measurements for the two ropes are given in Table 4. Bitting [6] has made measurements of the stiffness of 0.5 inch diameter double braided polyester ropes, but over a much more limited range of frequency, mean load and load range: compared on a basis of enclosed cross sectional area, the stiffnesses reported above are higher by a factor of between 2.8 and 4.

Creep and environmental assisted degradation

Assessment of the creep and environmental degradation was performed at two scales:

(i) industrial yarn (1100 dtex *Diolen*® 855TN from AKZO, measured breaking load: 87 N); and,

(ii) rope (approximate breaking load: 60 kN).

The yarn tested was from the same batch as used to make one of the ropes. The other rope was made of a similar yarn (*PET*® 855T from COBAFI) except for the absence of marine finish.

Yarn

Yarn samples of 100 mm length were tested under constant loads of 20%, 30% and 40% of their average breaking load for up to 1 year, in 3 environments: air, deionised water and synthetic sea water to ASTM D1141-75. Samples in water were kept under constant temperatures of: 4, 20, 30, and 40° C.

After fixed periods of time samples were removed and tested to assess any loss in tensile strength. A total of 99 valid results were obtained.

Although most of the data sets showed negative coefficients of correlation between breaking load and both time and temperature, the strength losses were not significant at a 5% level. For example, samples in deionised water at 4, 20 and 40° C showed average breaking loads of 85.75, 86.20 and 85.5 N respectively, all marginally lower than the new yarn strength of 87 N.

To check yarn elongation under load, 2 types of creep tests were performed: short term tests of up to one week and long term tests of up to one year duration. The former were performed on an Instron testing machine cycled between load limits set 0.02 N apart. The latter were under deadweight.

The results obtained agreed well with manufacturers' literature. They also agree with the findings of Tabor [7], that creep rates are nonlinear on a logarithmic time scale, as for a number of other yarns. Creep rates measured between 10^6 and 10^7 seconds are similar to those measured in aramid yarns: an elongation of 0.065% was measured between 10^6 and 10^7 seconds, at 30% UBL.

Rope

For both rope constructions, the combined effect of constant load and seawater was investigated. Creep loading was performed using four purpose-built machines in which the force of 30% ± 0.5% UBL was obtained by means of a constant area pneumatic actuator and two stage air pressure regulator. Elongation was calculated from measurements of the distance between sockets, corrected by measurements taken over a gauge length marked on the rope prior to sealing in a seawater bag. This method of extension measurement though far from being ideal, provided consistent results. Following exposure to creep loading, stiffness and breaking load were determined on a servo-hydraulic testing machine. The results obtained are shown in Table 5, in relation to the performance of new ropes.

Figure 4 shows the elongation results for 4 samples of Brascorda Parallel rope plotted against the log. time. To illustrate the significance of these results, the regression equation obtained can be used to work out the elongation of a 2500 m rope (as might be used for 1000 m water depth) at an average tension of 30% UBL. Between the end of the first day and the end of the first year such rope would stretch 15 m, but over the next 10 years it would only stretch a further 3 m.

Fatigue

Karnoski and Liu [8] report the only fatigue testing of polyester rope which is relevant to the aplication being considered here. They tested one sample of 2.5 inch diameter polyester *Superline®* rope between 8 and 25% UBL for 2 million cycles, finding a retained strength of 93% of UBL.

Most of the other reported studies of fatigue of fiber ropes are more relevant to loading typical of single point moorings. Work at NEL [9] on plaited and double braided constructions concentrated on much higher load ranges with rather low minimum loads. Some results have been reported for parallel laid ropes [10], but those also concentrated on higher load ranges.

Typical static tension levels for operating conditions are not expected to exceed 30% MBL for the West of Shetlands conditions, and 35% MBL for the Campos Basin. Significant wave frequency tension ranges for operating conditions are predicted to be less than ± 5% for West of Shetlands, and ± 2% in Campos Basin. In relation to the reported fatigue endurance of polyester ropes, these figures, even in isolation, virtually eliminate the likelihood of fatigue degradation. Testing targeted at developing relevant fatigue endurance data was considered to be completely unjustified. However some form of assurance that the particular ropes and their terminations would not be degraded by large numbers of cycles was thought necessary.

For each rope three tests were run for 1 million cycles at 20 ± 10% UBL, immersed in running tap water. This loading level whilst not expected to induce significant fatigue damage, is far in excess of what might be expected in the application. After cycling samples were removed from the test frame for examination and then replaced and, after a few further cycles to bed in the construction, loaded to failure. The values obtained are shown in Table 5.

Microscopic examination showed no degradation on the bulk of the rope samples, and only minor abrasion degradation in the yarn/resin interface region.

Conclusions & Recommendations

1 Practical, cost effective catenary mooring systems, using lightweight polyester ropes based on current technology can be designed for deep water operation. Such systems, whether using piles or anchors, provide attractive solutions in terms of station keeping performance, weight, MBL and fatigue, as well as cost.

2 In comparison with steel wire or chain, polyester ropes have a low axial stiffness that results in low loads amplitude at wave frequency. Test results showed no significant loss of strength after 1 million tensile fatigue cycles at load amplitudes typical of West of Shetlands storm conditions. Fatigue is therefore not considered to be a problem.

3 There are differences in axial stiffness whether loaded at first order wave frequency, low frequency or statically. These differences must be taken into account in designing the system.

4 The effective "installation" stiffness should also be considered when sizing the upper steel component in the type of mooring combination described here.

5 The use of low twist constructions gave high tensile efficiency, typically 75% for terminated ropes tested.

6 The potential problems of creep have been evaluated on the yarns and small ropes and do not appear significant. Operational creep deformation will be concentrated in early life and can be allowed for when sizing the upper steel component.

7 Results obtained to date indicate that long term exposure to a combination of constant tension and seawater are not damaging to polyester ropes over the typical time scale and load levels of a catenary moored FPS.

8 Before the implementation of designs for mooring systems based on polyester ropes it is considered necessary to perform full scale tests to check design data for stiffness, and to validate the performance of termination systems

9 Longer term future work should seek to:

(i) develop procedures for monitoring the degradation of polyester ropes;

(ii) investigate the bending-tension [11] performance of polyester ropes operating over sheaves, or other geometries of fairlead;

(iii) investigate low stiffness tethers for TLP's.

Acknowledgements

The authors gratefully acknowledge the provision of test samples by Akzo Fibers BV, Brascorda SA, and Marlow Ropes Ltd.

References

1 Reading Rope Research and Global Maritime *Lightweight Materials for Offshore Station Keeping* Final report of a joint industry study distributed by Reading University, 1989.

2 API *Design, Analysis and Maintenance of Moorings for Floating Production Systems* API RP 2FP1, draft issue for industry review, May, 1991.

3 Larsen, C. M. and Fylling, I. J. *Dynamic Behaviour of Mooring Lines* Proc. 'BOSS 82', at MIT. Edited by Chrys-sostomidis and Connor, Hemisphere Publ. Corp., August, 1982.

4 Salama, M. M. *Lightweight Materials for Mooring Lines of Deepwater Tension Leg Platforms* Marine Technology, No.3, 1984.

5 Walton, J. M. & Yeung, Y. C. T. *Flexible Tension Members from Composite Materials* 6th. ASME et al. OMAE, Houston, 1987.

6 Bitting, K. R. *Cyclic Tests of New and Aged Nyoln and Polyester Line* Proc. 12th OTC Conference, Houston, Texas, May 1980, paper no. OTC 3852.

7 Tabor, B. J. & Wagenmakers, J. C. *Some Longterm Behaviour Aspects of Industrial Yarn, Related to Environmental Influence* TechTextil Symposium, Frankfurt, May, 1991.

8 Karnoski, S. R. & Liu, F. C. *Tension and Bending Fatigue Test Results of Synthetic Ropes* Proc. 20th Annual OTC Conference, Houston, Texas, May 1988, paper no. OTC 5720.

9 Crawford, H. & McTernan, L. M. *Cyclic Testing of Continuously Wetted Synthetic Fiber Ropes* Proc. 15th Annual OTC Conference, Houston, Texas, May 1983, paper no. OTC 4635.

10 Crawford, H. & McTernan, L. M. *Fatigue Properties of Parafil* Symposium on Engineering Applications of Parafil Ropes, Impeial College of Science and Technology, London, January 1988.

11 Chaplin, C. R. *Bending-Tension Fatigue in Offshore Applications* OIPEEC (Organisation International pour l'Etude de l'Endurance des Cables) Bulletin No 52, December, 1986.

Table 1: Survey of properties of selected tethers

material	construction	strength kN/mm^2	modulus kN/mm^2	dry weight kg/mm^2 per 100m	wet weight kg/mm^2 per 100m	relative strength value	relative stiffness value
polyester	"Superline"	0.30	5.0	0.096	0.027	69%	14%
polyester	parallel	0.38	9.7	0.096	0.027	102%	32%
polyester	Brascorda //	0.47	9.0	0.096	0.027	77%	18%
HMPE-standard	braided	0.47	18.0	0.067	-0.002	31%	15%
Aramid-standard	wire rope	0.66	26.5	0.096	0.030	41%	21%
Aramid-standard	parallel	1.10	48.3	0.101	0.032	36%	20%
Aramid-hi.mod.	wire rope	0.66	41.6	0.096	0.030	41%	33%
Aramid-hi.mod.	parallel	1.10	76.7	0.101	0.032	31%	27%
steel wire (1770)	6x36 WS+IWRC	0.84	67.7	0.532	0.464	100%	100%
steel wire (1770)	Spiral Strand	1.02	108.9	0.633	0.532	77%	103%

NOTE: 1 Areas based on enclosed area, excluding any jacket.
2 Density units as mass per sq. mm cross section per 100 m length.
3 Strength and modulus as obtained from survey of manufacturers' literature
4 Strength value and stiffness value represent strength and modulus per $ in relation to six strand steel wire rope

Table 2: Summary of results of analysis for alternative mooring systems in Campos Basin environment

water depth	m	500	500	500	1000	1000	1000	2000	2000	2000
mooring system	–	poly+pile	poly+chain	wire+chain	poly+pile	poly+chain	wire+chain	poly+pile	poly+chain	wire+chain
chain diameter/length	mm/m	–	76/650	84/650	–	76/770	87/600	–	84/840	92/570
rope diameter/length	mm/m	115/1250	112/1250	96/1000	116/2500	116/2500	103/2250	117/5000	125/5000	115/4000
chain MBL	kN	–	4620	5550	–	4620	5920	–	5550	6540
rope MBL	kN	4620	4380	5740	4700	4700	6520	5460	5460	7910
mean tension	kN	1993	1983	2350	2280	2299	3005	2815	2865	4049
low frequency tension	kN	478	408	268	416	339	252	407	299	206
wave frequency tension	kN	202	173	715 *	104	107	585 *	62	68	513 *
sig. wave freq. tension range	% MBL	8.8	7.4	13.2	4.4	4.6	9.6	2.28	2.5	7.0
maximum tension	kN	2673	2564	3333	2800	2745	3842	3284	3232	4768
maximum tension	% MBL	58	59	58	60	58	59	60	59	60
wave freq tension/EAε	–	0.89	0.81	0.46	0.90	0.93	0.73	1.06	1.02	0.91
maximum offset	m	30.8	46.4	47.3	64.8	88.1	83.9	148.7	163.5	205.0
maximum offset	% WD	6.2	9.3	9.5	6.5	8.8	8.4	7.4	8.2	10.3
capital cost for 8 lines	k US$	756	1799	2052	1542	2822	3203	3582	5260	5493
total vertical force	kN	2832	2680	5288	3312	3152	7960	4104	4144	15376

* NOTE: maximum tension dominated by wave frequency components in these cases

Table 3: Summary of results of analysis for alternative mooring systems in West of Shetlands environment

water depth	m	500	500	500	1000	1000	1000	1000	2000	2000	2000
mooring system	–	poly+pile	poly+chain	wire+chain	poly+pile	poly+chain	aramid+chain	wire+chain	poly+pile	poly+chain	wire+chain
chain diameter/length	mm/m	–	100/400	111/525	–	84/600	92/780	107/600	–	84/720	107/280
rope diameter/length	mm/m	150/1500	145/1500	122/750	125/2500	125/2500	94/1500	122/1500	125/5000	125/5000	128/4000
chain MBL	kN	–	7600	9130	–	5550	6550	8560	–	5550	8560
rope MBL	kN	7860	7340	8970	5460	5460	6625	8970	5460	5460	9780
mean tension	kN	2095	2377	2599	2124	2177	2166	2854	2449	2424	3893
sig. low frequency tension*	kN	373	273	304	300	215	274	255	231	194	188
wave frequency tension	kN	2004	1455	2265	830	718	1488	1964	424	402	1644
sig. wave freq. tension range	% MBL	27.4	21.3	27.2	16.4	14.2	24.2	23.6	8.4	8.0	18.0
maximum tension	kN	4472	4105	5168	3250	3110	3878	5073	3103	3020	5729
maximum tension	% MBL	57	56	58	60	57	59	57	57	55	59
wave freq tension/EAε	–	1.00	0.74	0.19	0.95	0.82	0.40	0.34	0.97	0.92	0.68
maximum offset	m	27.6	33.5	47.1	56.6	68.8	97.4	98.0	112.5	131.0	190.7
maximum offset	% WD	5.5	6.7	9.4	5.7	6.9	9.8	9.8	5.6	6.6	9.5
capital cost for 8 lines	k US$	1547	2539	2608	1791	2990	5453	3614	3582	5020	5995
total vertical force	kN	3224	3893	8054	3366	3688	3415	9916	4104	4147	18056

* NOTE: maximum tension dominated by wave frequency components in all cases

Table 4: Rope stiffness measurements

Loading Regime	*Superline* ® *EA* (kN)	Brascorda // *EA* (kN)
"Instalation"	300	370
10 ± 5% of UBL - 7.5s period	1043	1324
10 ± 5% of UBL - 200s period	990	1246
50 ± 2.5% of UBL - 15s period	1669	1923
50 ± 15% of UBL - 15s period	1416	1698
20 ± 10% of UBL - 15s period	1059	1290
20 ± 10% of UBL - 1 week period	–	1072

Table 5: Rope strength measurements

Condition	*Superline* ® mean breaking load (kN)	Brascorda Parallel mean breaking load (kN)
new	57.2 std. dev. 1.7	58.7 –
after creep at 30% UBL	53.8 (94% of new) std. dev. 4.2	58.1 (99% of new) std. dev. 2.8
after fatigue at 20±10% UBL for 10^6 cycles	56.0 (98% of new) std. dev. 3.3	55.8 (95% of new) std. dev. 4.6

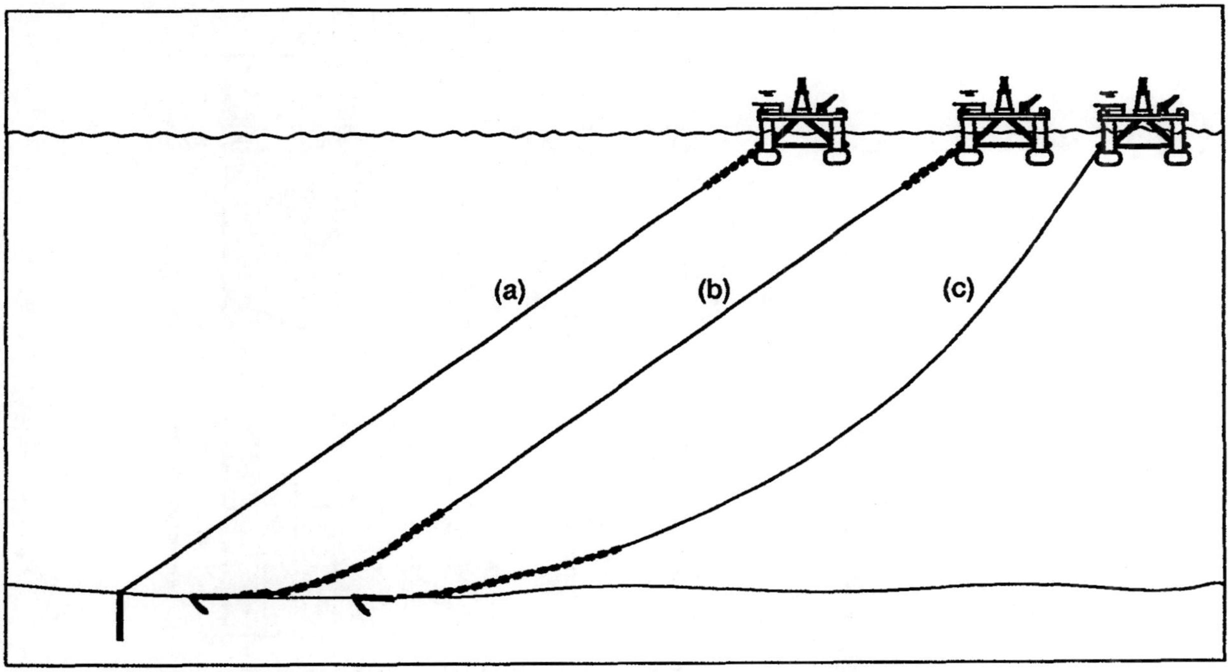

Figure 1: FPS with three mooring systems considered in the analysis: (a) lightweight rope & pile, (b) lightweight rope & anchor, (c) wire rope & anchor.

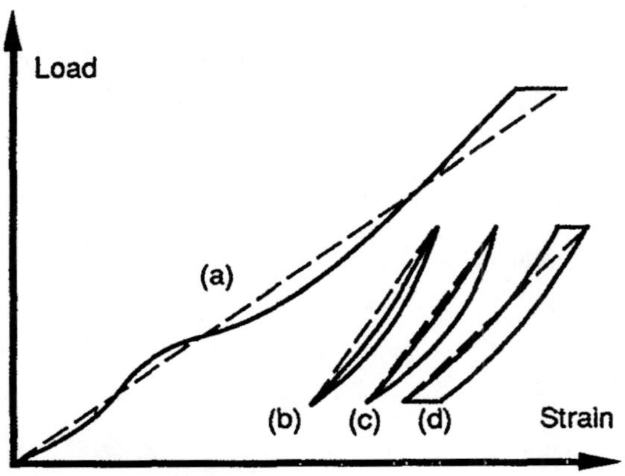

Figure 2: Definitions of stiffness for (a) "installation" with hold at the maximum load, (b) wave frequency, (c) low frequency, and (d) "static"

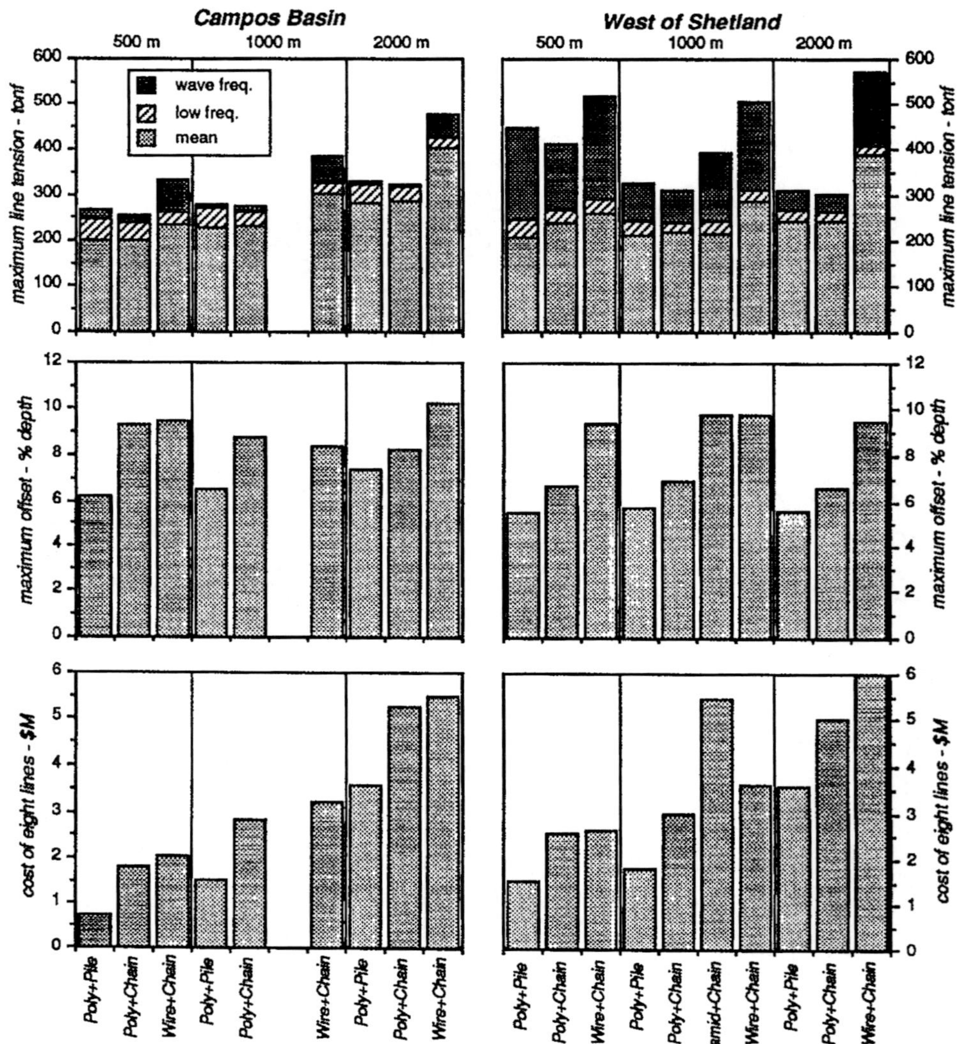

Figure 3: The maximum line tensions (top), maximum offsets (centre) and mooring line costs, calculated for FPS catenary moorings located in Campos Basin (left) and West of Shetlands (right) in 500, 1000 and 2000 m water depth, for different line combinations.

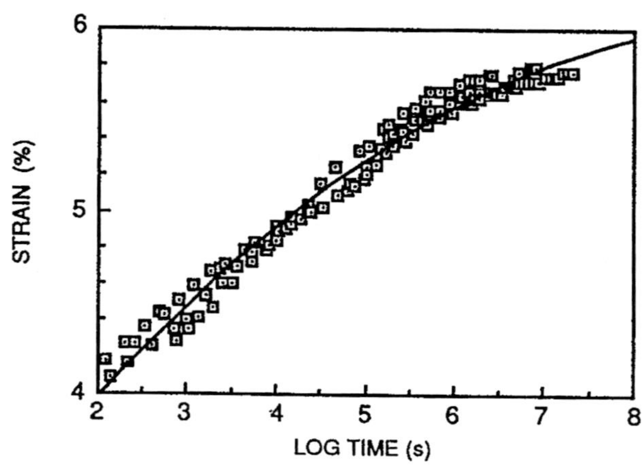

OTC 8452

Heave Compensated Landing System - A Novel Tool for Subsea Intervention

Ron Nelson, SPE, Kerr-McGee Corporation, and James Soliah, Oceaneering International, Inc., and Trevor Caldwell, BP Exploration Inc., and Denby Morrison, Shell E&P Technology Company, and John Pritchard, Shell Offshore Inc.

Abstract

This paper presents a new intervention system that allows the use of a low cost workboat to deploy and recover delicate payloads in subsea installations. Vessel motions are decoupled from the subsea package by a specific chain and buoy system. A prototype was built and successfully tested offshore in 3,000 ft of water during May 1996. The Heave Compensated Landing System (HCLS) concept, offshore testing program, test results, lessons learned, and future developments are discussed.

Introduction

Most subsea production systems are equipped with smaller components designed to be recovered and replaced using less expensive, non-invasive intervention techniques. These components include subsea control pods, specially designed valve and choke trim and actuators, and fluid distribution modules. They are typically designed to be recovered by a Remotely Operated Vehicle (ROV) intervention system operated from a support vessel. The subsea components usually require "soft landing" on the manifold because of delicate components or interfaces.

Because of the need for a "soft landing," the deployment system is usually mounted on a large, stable vessel such as a semi-submersible drilling rig or derrick barge. Smaller workboats are rarely used because their heave motion, even in modest seas, poses significant risk to the subsea equipment during landing or recovery operations. Unfortunately, the high cost and questionable availability of large offshore vessels may prohibit their use.

This system provides a cost-effective, workboat based alternative for servicing subsea facilities. The payload is not supported directly by the vessel, but is instead supported by a buoy below the wave zone. The buoy is controlled by a chain and wire rope linking it to the workboat. The result is a stable, inexpensive, maneuverable system capable of deploying and recovering subsea payloads in a wide range of water depths. No special tooling is required.

System Description

Concept Overview. The HCLS is a passive system that requires no active damping systems. The concept originated from early pipeline intervention methods[1]. Nearly all the components are readily available, and most can be leased.

The HCLS decouples vessel motion from the payload by supporting the payload from a buoy below the wave zone. See Figure 1. Because the buoy is below the wave action and its associated turbulence, there is little energy and therefore little tendency for inherent motion.

The buoy is connected to the vessel by chain and wire. The short chain section (~400 ft) is attached to the buoy and hangs to form a "belly" before rising toward the vessel.

The chain serves many purposes. The chain "belly" allows the workboat to heave independently of the buoy. Also, because the chain's weight is supported by both the buoy and the vessel, the buoy will naturally come to equilibrium with the sum of its buoyancy, payload and partial chain weight. Thus, the chain automatically facilitates trim adjustment for small weight inaccuracies.

The chain is also used to transfer weight to or from the buoy after the payload is landed or recovered. For example, consider the case of a 4,000 lb. control pod being lowered and landed on a subsea manifold. After setting the pod on the manifold, 4,000 lb. of chain weight (~105 ft of 2 inch chain) can be transferred to the buoy by paying out additional cable from the deployment system. As more chain weight is transferred to the buoy's side of the chain "belly," the equivalent payload weight is transferred to the subsea landing base. The payload can be released from the running tool, and the system will remain in equilibrium.

As the vessel stern heaves up and down, the neutral point in the chain "belly" shifts and transfers chain weight to and from the buoy. This load transfer could theoretically cause the buoy to move up and down, defeating the purpose of the HCLS system. However, this type of motion can be eliminated by engineering the HCLS around the resonant periods of each sub-system. When properly designed, the chain load is transferred to and from the buoy too quickly for the buoy/payload package to respond. This effectively decouples the buoy from the vessel. Specific attention must be paid to the environmental conditions.

Another possible function of the chain is to eliminate "snap loading" on the deployment wire under overboarding or retrieval operations. Under some conditions, especially when line loads are light or payload sail area is large, the subsea system could be traveling upward while the vessel stern is moving downward. This transition from tensile to a zero load, then back to a tensile load is known as a "snap" load. These loads can damage or break the wire and/or rigging. Increasing the load at the end of the wire with an added chain section can assist in keeping the wire in tension at all times, and thereby reduce the snap load tendency.

Components. The following components make up the core of the HCLS system, as was successfully tested offshore in the Gulf of Mexico:

Running Tool / Payload. This is a job-specific piece of equipment. This is the item that is to be deployed to, or recovered from, the subsea installation.

Pendant Line. This links the running tool to the buoy and will typically be a segment of wire rope. It must be strong enough to support the subsea payload, running tool, and its own line weight with a significant safety margin to allow for wear and/or dynamic loads. If the payload area is large, hydrodynamic drag from its "sail area" should also be included. This is especially critical during deployment and recovery operations, as the line will be short (resulting in a very stiff system) and the payload will not be compensated.

Buoy. The buoy supports the running tool, payload, pendant line, and part of the chain weight. It operates below the wave zone and ideally below the surface current. The required buoyancy is the sum of payload weight, running tool and associated rigging weight, pendant wire weight, submergence allowance, and trim allowance chain weight.

Chain. The chain is connected to the buoy and forms a "belly" before turning up to the vessel deployment line.

Selecting the chain size and weight requires establishing a balance between optimizing the chain "belly" below the buoy and decoupling the buoy from the boat. The chain size should facilitate a reasonable "belly" length, be easily handled on deck, and be fairly light. A very heavy chain will transfer more weight to and from the buoy as the boat heaves and could possibly induce motion in the buoy. We chose 2 inch, 37.3 lb/ft chain for the sea trials.

The chain length is derived from considering four components: submergence allowance, trim allowance, payload weight allowance, and snap load allowance.

The submergence allowance is the chain weight that, with the sum of the payload and wire weights, is required to sink the buoy. This length also contributes to the allowable "travel" for vessel heave. Under our test conditions, the minimum chain length should be about 30 feet. Very long lengths are undesirable, as they can increase the risk of entanglement.

The trim allowance is an excess length to allow for buoyancy variations. For example, with a 6,000 lb payload and running tool and 1,500 ft of 1¼" wire, the total weight is 11,700 lbs. If 2 inch chain is used, the trim allowance could be 15-30 ft. This figure can be adjusted, depending on weight, buoyancy, and vessel heave calculations.

The payload weight allowance is simply the payload weight (exclusive of the running tool) divided by the chain's weight per foot. For example, a 4,000 lb payload would need 105 ft of 2 inch chain in the payload weight allowance section.

The snap load allowance will vary considerably with different HCLS configurations and sea-states. For our sea trials, we calculated that 17,000 lbs was required to overcome the snap load tendencies. This equated to 225 ft of 2 inch chain. One should also vary this length to be sure that the end of the deployment wire is never below the buoy, as this could pose some risk of entanglement.

Deployment Line. This connects the chain to the vessel-mounted deployment system and guides the subsea equipment. The deployment line must be strong enough to support the chain weight and its own line weight with a significant safety margin to allow for wear or shock loads. Also, one must consider hydrodynamic drag from the buoy in a worst case where the chain is entangled with the buoy and the system is uncompensated.

Swivels. Swivels are required to compensate for wire rotation. Standard wire rope is not torque-balanced and will twist as load is applied and relaxed. In the case of the HCLS, which employs thousands of feet of wire, this can cause twisting and entanglement of the subsea equipment. Torque-balanced wire is available but is expensive and usually not 100% balanced. Swivels placed into select points allow the wire to react without entangling the system. Ball bearing swivels are required because of their low turning friction.

A-Frame. The A-frame is used to overboard the HCLS equipment. Individual components are launched by the A-frame supported overboarding line and carried over the stern of the vessel. Requirements include sufficient dynamic load rating, sufficient clearance for the HCLS components, and sufficient travel to pick up the components on deck and swing them clear of the stern.

Overboarding Line. This line is used with the A-frame to deploy the HCLS components. The overboarding line passes through the A-frame sheave and lifts the equipment off the deck. The A-frame then booms out over the stern, and the tool is lowered below the surface. Once submerged, the load is transferred from the overboarding line to the deployment line. An ROV then releases the overboarding line. The operation can be repeated for the each component. The overboarding line must be strong enough to lift the heaviest component (usually the buoy) and tolerate snap loads with the buoy near the surface. These snap loads can be very severe and are case specific.

Winch. A double drum unit can be used to support both the overboarding and deployment lines. The system requires a large drum capacity to handle large amounts of wire and chain, and high speed to transit to and from the seafloor. Anchor-handling winches generally meet these requirements.

Component Rigging / Slings. As with the overboarding line, the component rigging must be strong enough to lift the heaviest component and tolerate snap loads with the buoy near the surface. As a result, the rigging lines and shackles should be oversized. Also, recovery loops on the buoy and running tool should be large enough to facilitate ROV recovery while the load is uncompensated.

Air Tuggers. Two air tuggers can be deck mounted on either side of the winch. They are used to place loads on deck and to help control loads during transfer to overboarding position beneath the vessel's A-frame.

Deck Crane. Used to move and organize equipment on deck.

Remotely Operated Vehicle (ROV). The ROV provides visual feedback to the operators, final guidance of the payload, and operates the recovery tool's latch / release mechanism.

Workboat. The workboat provides the support system for the HCLS and its crew. Requirements include an adequate A-frame, sufficient room to perform the overboarding operations safely, and a dynamic positioning and navigation system capable of locating and holding station over the work site.

Analytical Modeling of In-Place and Deployment Responses

Substantial numerical modeling and design of the HCLS configuration were undertaken before the offshore sea trials. These calculations supported design decisions in two main areas:

(1) The desired "In-Place" behavior and required sizes, weights, buoyancy, lengths and capacity of submerged (lines, chains, buoys) and surface-support (A-frame) components.

(2) The possible behavior of the configuration during deployment and retrieval of the major components.

In-Place Response. Single-degree-of-freedom and two-degree-of-freedom representations were made of the deployed system. These simple models were especially useful because they allowed for scanning of many probable seastates in the Gulf of Mexico. A later discussion will support the simple models by modeling the system with discrete finite elements. The calculations predicted that, for most practical systems in up to 3,000 ft water depths, the system's dynamic response will be relatively benign. An example of calculated response is shown in Figure 2. However, in deeper water (~6,000 ft) the vertical motion of the "chain" region could be amplified, resulting in very large heave relative to surface motions.

A Simple Model used to Scan for Possible System Responses in the Gulf of Mexico. The deployment line and chains can be approximated as a single-degree-of-freedom system. The stiffness, k_{line}, of the system can be derived from the axial wire properties and the length of the deployment line:

$$k_{line} = A_{line} E_{line} / L_{line} \quad (1)$$

where: A_{line} = effective line area;
E_{line} = effective axial modulus (a function of the line type, often around 13,000 ksi);
L_{line} = line length.

The mass is approximated as the weight of the chains at the bottom of the deployment line, and the effective mass of the deployment line which is taken as 1/3 the weight of the line. An allowance is made for added mass contributions. The simple model natural period is:

$$T_n = (M_{effectve} / k_{line})^{0.5} 2\pi \text{ (seconds)} \quad (2)$$

where: $M_{effectve}$ = effective mass;
k_{line} = effective axial spring.

One of the main variables in this subsystem is the length of the deployment line, which gets longer as the intervention depth increases. For similar packages this will be the cause of significantly longer natural periods of response as the system depth increases. The majority of practical configurations in water depths up to 3,000 ft will have periods in the 3 to 4 second range. The 6,000 ft water depth systems could have periods from 4 to 5 seconds.

It is important to evaluate the likely response of the range of systems in the relevant environmental conditions. Planning intervention during different times of the year will require seasonal environmental information. For example, by simply assuming annual data and not distinguishing between various seasonal differences, it can be deduced[2] that a seastate with H_s of 4 ft and T_p of 6 seconds (where H_s = significant wave height and T_p = peak period) occurs most frequently - about 10% of the time. This type of environmental data was used to scan the relative responses of "soft" systems (representative of 6,000 ft water depths) and "stiffer" systems (representative of 3,000 ft water depths)[3]. The models indicated that the chain region in a "soft system" is amplified relative to the stern motion. For example, if we look at a seastate represented by H_s of 6 ft and T_p of 6 seconds, expected maximum chain amplitudes of between 35 and 40 feet are possible in a 3-hour storm. In a 10 ft sea, which occurs only 0.5% of the time, the chain amplitudes can approach 50 ft. The maximum chain heave amplitudes are estimated to be 25-30 ft in the most frequently occurring storms.

Detailed Finite Element Models. Beyond modeling the HCLS as a single degree of freedom system, a finite element model of several hundred elements was constructed. The time-domain analyses[4] were conducted to confirm trends seen in the simplified models.

Many conclusions were made from the finite element analyses:

- The solutions are often difficult to obtain because of the physical properties of the configuration. For example, the chain sections tend to hang immediately adjacent to the pendant line but can be separated with a very small lateral load of a few pounds at the buoy location. This is similar to observations in the pipeline intervention systems[5,6] that showed an inclination for line fouling.
- Lateral isolation of the buoy from surge motions at the waterline is very good, with lateral buoy motions of fractions of an inch.
- Stiff ($T_n \approx 3$ seconds) systems experienced very little amplification of motions from the stern heave to the heave at the chain locations, whereas the soft systems ($T_n \approx 4.5$ seconds) could easily be forced into "snaploads" (amplitude of motion causing loss of tension) in the deployment line.

Predicted and Measured In-Place Behavior. The numerical predictions were successful in both sizing and configuring the system and predicting in-place response.

The physical tests showed that the calculations used to choose line sizes and lengths, chain weights and lengths, and buoyancy module were adequate.

Little amplification at the chain region of the tested (3,000 ft) configuration was expected because these were relatively "stiff" systems. Additionally, the tested configurations were calculated to have little motion at the landing package.

Seas of about 3.5 ft were measured during the sea trials. Maximum stern motions of more than 6 ft were measured[7], and the chain region experienced similar magnitudes of vertical heave without discernible amplification. Motions at the isolated landing package could not be detected, to an accuracy of fractions of an inch, even under the roughest test conditions.

Attempts to test the behavior of softer systems offshore were not completed. However, the added confidence gained with the numerical tools could be used to extrapolate expected behavior in other yet untried systems.

Different damping models were assumed before the physical tests in order to predict maneuverability[8]. The tested configurations with a single large buoy and substantial chain lengths confirmed that good, damped characteristics were achievable.

Modeling of Deployment and Retrieval Response. Simple calculations were made to estimate possible snaploads during the deployment and retrieval of the system through the wave zone. This was used to help evaluate the surface-support and submerged components before mobilization.

Spreadsheets were used to calculate the line tensions and displacements at critical locations for the buoy retrieval with the overboarding line[9]. The output includes the possible lack of line tension at the bottom of the deployment line.

A harmonic stern motion and constant winch speed were assumed as input to the spreadsheet. The periodic loss of overboarding line tension caused "impulsive" loads, which in turn could cause relatively large (80,000 lb) transient loads. Analyses revealed that the peak loads were caused by transients associated with a very long (10's of seconds) periodic input. Once the transients were overcome, the peak line loads were often less than one-half of the transient loads. Owing to the randomness of actual sea states, the periodicity (and thus the calculated peak loads) assumed in the calculations may be difficult to achieve in nature.

Offshore Test Program

The project partners (BP Exploration, BHP Petroleum, Exxon USA, Kerr-McGee Corporation, Shell E&P Technology, and Shell Offshore) agreed to share the costs of an offshore system demonstration. Oceaneering International was contracted to provide engineering and project management. Lazer Inc. was subcontracted to provide rigging services and equipment.

Test Objectives. The offshore test was intended to prove the HCLS concept and gather data on handling characteristics and system responses. Objectives included:

- Prove that the buoy-supported payload would be effectively decoupled from the vessel motion.
- Test a number of buoy and chain configurations and determine which were best suited to offshore operations.
- Verify equipment handling and estimate overboarding procedures and requirements.
- Test the system's sensitivity to factors such as vessel orientation and sea state.
- Test dynamic properties such as lateral pendulum periods, vertical overshooting in response to a sudden stop, and hydrodynamic drag.
- Verify ROV operability, including the ROV's ability to land the system when the vessel and buoy is laterally offset from the target.
- Test the vessel's ability to maneuver the system around the landing base without ROV intervention.
- Test different payload landing and lifting techniques.
- Gather sufficient data to refine the analytical model further.

- Determine workboat needs such as deck space, dynamic positioning, sea state response, etc.

Test Overview. The test was intended to prove the HCLS concept, not to build a working tool. Simulated equipment was used for the running tool, payloads and landing base. This reduced test costs and risk to expensive subsea tooling. The following components made up the HCLS offshore test configuration:

Running Tool. The running tool was a "typical" tophat-type design with a downward looking funnel for stabbing alignment. Dry weight was 1,710 lbs.

Payloads. The payload was designed to be run in one or two segments, depending on test requirements. Module #1 weighed 2,075 lbs and module #2 weighed 2,085 lbs dry.

Landing Base. The landing base was a mock-up of a subsea manifold with a valve body. The payloads were landed from above through a square 3 ft x 3 ft access area, with no guidance funnel.

Pendant Line. The pendant line was a section of 1¼" 6x25 EIPS wire rope that could be varied from 450 ft to 1,450 ft in length.

Buoy. The syntactic foam buoy was 13 ft tall and 8 ft in diameter and weighed 12,700 lb dry. It provided 10,605 lbs of buoyancy in seawater. It was surrounded by a 1,700 lb protective cage to prevent chafing from the chain motion.

Chain. A 540 ft section of 2 inch chain with a dry weight of 37.3 lb/ft was used.

Deployment Line. A 4,000 ft length of 1¼" 6x25 EIPS wire was installed on the winch's lower drum.

Overboarding Line. A 500 ft length of 1¼" 6x25 EIPS deployment wire was installed on the winch's upper drum.

A-Frame. A 25 ton hydraulically operated A-frame was built and permanently installed on the workboat for the test.

Winch. An Intercon Model DW-200 Towing / Anchor handling winch was used for the test. Line capacity was 3,000 ft of 2 inch wire rope per drum. Line pull was 350,000 lb nominally.

Workboat. Oceaneering's ROV support vessel "M/V Ocean Service" was contracted for the offshore tests. Length is 200 ft with a deadweight of 1,434 tons. The vessel is equipped with a 25 ton stern-mounted A-frame, a midship launched ROV, a 35 ton pedestal-mounted crane, a Nautronix Dynamic Positioning system, and quarters for 32 personnel.

Instrumentation. A critical objective of the test was to record system response to varying environmental conditions. This data is being used to calibrate existing mathematical models. Highly accurate instrumentation with high-frequency data recording was used during the test. This included two tri-axial vessel motion sensors; deck-mounted deployment line tension, speed, and payout meters; an ROV cage mounted current meter; a wave-rider buoy; and an acoustic tracking system to record ROV location relative to the vessel. Most of this data was recorded in real-time on a computer-based system.

Test Location. The HCLS sea trials were conducted in Garden Banks Block 470, about three miles south of Shell Offshore's "Auger" facility. Water depth was ~3,000 ft. This site was chosen because Shell had performed a detailed bottom survey showing no hazards, the water depth matched desired test scenarios, and the Auger facility has equipment capable of measuring and recording various environmental conditions. This included an acoustic doppler current meter, wave-rider buoy, and wind meter.

Test Results. Three different system configurations were tested. A description and summary of each system follows.

Test #1 - "Three Chain" Design. We considered this our "base case" design prior to offshore testing. In addition to the normal 540 ft chain section, a 40 ft section of 2" chain was attached parallel to the main chain to double the weight per foot at the base of the vessel side of the chain belly. We felt that by having a heavy chain section near the belly, one could more readily transfer payload weight after the payload was landed on the test base. It also reduced the length of the "belly" hanging below the buoy after the payload was released from the recovery tool. This was believed to reduce entanglement potential.

Unfortunately, the flat shoulder created by the doubled chain section proved to be a significant snag point during overboarding operations. The chain and pendant line became entangled and could not be cleared by the ROV. This configuration had to be recovered and was not re-attempted. This HCLS configuration requires very accurate weight estimates and careful rigging during deployment.

Test #2 - "Single Chain with Deep Buoyancy" Design. This case employed a single 540 ft chain section and a short (450 ft) pendant line. The buoy submerged to operating depth at 2,500 ft.

This configuration worked very well. It was easily deployed and effectively decoupled vessel motions from the payload. It was also easily maneuvered by the surface vessel and easily manipulated by the ROV.

Specifically:

- The vessel stern motion was measured at ±6 ft of heave, while no motion was detected at the payload.
- Payload weight was varied between 2,000 and 4,000 lbs. No significant performance changes were detected.
- Vessel orientation to the prevailing weather affected vessel heave but no effect on the subsea payload motion.
- The system displayed exceptional motion "damping." A pendulum was created by laterally offsetting the payload from the buoy by 50 ft. It was then released and the system returned to equilibrium without detectable oscillation. Likewise, the buoy overshot its equilibrium point by 6-10 ft when the winch was stopped suddenly during high-speed transit, but it quickly stabilized with few, if any, oscillations.

- The 75 HP ROV easily controlled the payload and tool. It could swing the tool laterally out to 100+ ft away from the buoy, and raise and lower the payload during the soft landing / recovery without the aid of the winch.
- The ROV could land the payload when the vessel was offset from the landing base by 100+ ft horizontally.
- The system maintained its position relative to the vessel when the vessel was maneuvered around the landing base. We were able to move several hundred feet off location, then readily reposition the tools above the manifold without ROV input.
- The system can be easily maneuvered by the vessel. HCLS components can be overboarded in a safe area away from the subsea facilities, then positioned for landing using the vessel's dynamic positioning system.

This test and configuration were highly successful and far simpler than the configuration employed on Test #1.

Test #3 - "Single Chain with Near Surface Buoyancy" Design. This case employed a single 360 ft chain section. The pendant line was longer (1,450 ft), potentially allowing the use of a shallower, less expensive buoy, or to extend the reach of the system beyond 3,000 ft.

This system also worked very well. Like Test #2, it was easily deployed and effectively decoupled vessel heave motion from the payload. It was also easily maneuvered by the surface vessel and easily manipulated by the ROV. Its performance was virtually identical to Test #2 but required a potentially far less expensive buoy. The main drawback is that it required significantly more time for the ROV to survey the system and verify that there was no entanglement after deployment or payload release / recovery. The survey time was much greater because of the increased distance from the running tool (where the ROV guides the landing / recovery operations) to the chain "belly" (the most likely place for entanglement).

HCLS System Deployment Procedure Overview

Following is a summary of the HCLS deployment procedures used during the offshore testing.

Mobilization. During mobilization, the winch should be placed to the bow end of the work deck. The A-frame should be at the vessel stern. The running tool should be below the A-frame and ready for deployment. The buoy, chain, etc., should be stowed on deck.

Overboard Running Tool. *(See Figures 3, 4 and Photo 1)* The running tool is deployed first.

- Two lines are attached to the running tool rigging: the pendant line, which is linked to the "deployment line" on the winch's lower drum and runs over the stern roller, and the "overboarding line," which is on the winch's upper drum is routed through the A-frame sheave. The "deployment line" is slack, and the "overboarding line" is used to lift the payload off the deck and carry it over the stern as the A-frame is boomed out. The "overboarding line" is connected to the running tool rigging by a ROV-operable shackle. The air tuggers are connected to the running tool by softline loops. By applying tension with the tuggers, the load is controlled as the vessel pitches and rolls.
- After booming clear of the stern, the softline loops are cut to free the tool from the air tuggers. The tool is lowered to ~400 ft where an ROV is waiting. The "deployment line" is raised until it is tensioned (carrying the load) and the "overboarding line" slack. The ROV then operates the ROV shackle, which releases the "overboarding line" from the running tool. The "overboarding line" is then recovered to surface.
- The running tool is lowered until the socket linking the pendant line to the "deployment" line is on deck. The pendant line socket is then set in a "pelican hook," which secures it to the deck. The load is then transferred to the pelican hook by paying out cable until the overboarding line is slack on deck. It is then released from the pendant line.

Prepare the Chain. The chain is loaded onto the winch. Ideally, the chain should be loaded on the bottom drum, but few winches can hold 4,000 feet of 1½ inch wire and 400 feet of 2 inch chain on one drum. Because of this, the upper drum is temporarily used as the "deployment" system and the lower drum as the "overboarding system."

- Remove the "overboarding line" from the A-frame sheave. Secure one end of the chain to the "overboarding line" in the top drum. Load all the chain onto the drum. Then pay the chain out over the stern and allow it to untwist. Reel it back onto the drum.

Launch the Buoy. *(See Figure 5 and Photo 2)* It is important for the buoy to clear the stern area as quickly as possible, as the props and rudders could be damaged if the buoy is swept under the vessel. Also, the A-frame should not carry the buoy load after it is submerged, as its large sail area can create tremendous loads that will be reacted by the A-frame.

- Run the lower drum line over the A-frame sheave. This will temporarily be used as the "overboarding" line.
- Attach a "tri-plate" to the base of the buoy with a 10 ft segment of wire rope. Attach the pendant line to one side of the tri-plate, and the chain on the top winch drum to the other side of the tri-plate. The top of the tri-plate is linked to the base of the buoy by a 10 ft segment of wire rope. Attach the "overboarding line" to the top of the buoy with an ROV operable shackle.
- Pick the buoy off the deck with the "overboarding line" (bottom drum). Reel in the chain "deployment line" (top drum) and release the pendant line from the pelican hook.

- ➢ If possible, shut down the vessel propulsion system to eliminate the risk of the buoy drifting or being sucked into the props while they are rotating. The propulsion system can be returned to operation when the buoy has cleared the stern.
- ➢ Launch the buoy over the stern. The running tool will be hanging from the tri-plate below the buoy. Lower the buoy into the water. The buoy will not sink until sufficient chain is hanging in the "belly" below the tri-plate. As soon as the buoy is floating, pay out extra "overboarding line" over the A-frame to prevent a sudden heave from shock loading the A-frame. Pay out chain until the buoy submerges and clears the vessel stern. Keep the "overboarding line" slack, being careful not to entangle the line with the "deployment line."

Final Deployment. *(See Figure 6)*

- ➢ When the chain is fully paid out, secure the end of it to the deck with a pelican hook. Slack the upper drum line and disconnect the upper drum wire from the chain. Have the ROV operate the shackle and release the overboarding line (lower drum) from the buoy. Recover the line to surface.
- ➢ Remove the lower drum wire from the A-frame sheave and attach it to the chain. Tension the line and release the chain from the pelican hook. The HCLS is now fully deployed and ready for lowering.

The HCLS can be recovered by reversing the deployment operations. A hook with ROV-operable lock can be used to connect the loads to the overboarding line.

Lessons Learned

The driving force behind the offshore testing was to learn about the system behavior and deployment operations. Many lessons were learned, especially about the rigging and deployment. These are summarized below.

Entanglement. This was and will remain a major issue. As learned while deploying the first test configuration, the system can become hopelessly twisted and knotted if proper precautions are not taken.

Wire. As the finite element models predicted, the chain and pendant wire hang immediately adjacent to each other below the buoy. Most wire rope is not torque-balanced and will twist as load is applied and released. In the case of the HCLS, this can and did cause twisting and entanglement of wire and chain. For example, after releasing the 4,000 lb payload and picking-up the running tool, the running tool was observed to rotate more than 30 times as the torque was released from the wire. Torque-balanced wire is available but is expensive, and it is usually not 100% balanced. Simple plate-type swivels were used in the system, but they had too much turning friction and proved useless.

Ball-Bearing Swivels. It is strongly recommended that ball bearing swivels be used to allow the wire to react its torque without entangling the system. Ball bearing swivels are required instead of cheaper designs because of their low turning friction.

Tool and Equipment "Shoulders." The upper section of the pendant wire hangs parallel to and very near the chain. This creates a significant risk of entanglement. Care should be taken to eliminate any shoulders or obstructions that could snag or entangle the wire and chain. This includes wire sockets, shackles, and swivels.

Rigging Equipment.

Slings and Shackles. The sea trials did not measure loads during deployment and retrieval. However, it was very apparent offshore that the snap loads could be large relative to the fully deployed system response loads, and that these do warrant careful attention.

Oblong Links. Because of its high strength and low cost when compared with a round link, an oblong link was used as the connection point between the running tool rigging and the overboarding and deployment lines. The oblong link was too narrow and caused interference between the two lines. In one case, the two became jammed at the top of the link and had to be recovered to surface to be separated. We feel that an oblong link is well suited to carry one line, but should never be used to carry two lines.

ROV-Operated Spring Shackles. Extreme caution should be used when these devices share a connection point with another shackle. They can became entangled, or the ROV release pin on the spring shackle can be accidentally tripped by the other shackle. This can create operational problems and a potential safety hazard. We feel that a conventional shackle modified with an ROV "tee" handle would provide a better connection mechanism.

Spare Shackles. Shackles are used for a variety of operations, and many spares of all sizes should be on location. Uses include serving as a winch wire "auto-wind" guide when used in conjunction with the air tuggers, as improvised sheaves for the air tugger lines on deck, and as spares if any are lost over the stern during deployment or recovery operations.

Stern Roller. A stern roller is essential to reduce deployment line wear over the vessel stern. It should have a large enough diameter to allow the chain to pass without hanging up on the link connections. Vertical side rollers would be useful, as surface currents can pull the buoy and wire off the stern roller center.

ROV Operations. Many different methods of landing and recovering the payload from the landing base were tested. Criteria for success were based on soft landing and payload controllability. Generally, performing either operation was very easy.

Payload Landing. The preferred method to land the payload was by using the vessel and winch to spot the payload immediately above its target. Once above the target, the ROV was used to pull the payload down to its landing base. This operation was very controllable and was generally "fail safe." The winch was locked, and any ROV system failure would result in the payload being released, allowing it to return to its neutral point above the manifold slowly where there was no danger of damage. Once the payload was on its landing base, the winch would pay out the required chain weight and the payload would be released.

Payload Recovery. The preferred recovery method was similar to the landing method. After the payload was secured to the recovery tool, the ROV grabbed the tool and maintained 50% downward thrust. The winch operator then started reeling in sufficient chain weight to lift the tool, payload, and ROV. When the ROV and tool started moving, the ROV stopped thrusting and rode with the tool until it cleared the landing base area. By employing this technique, the ROV was able to maintain control over the tool at all times. In case of a winch or ROV failure, the system's "neutral point" without ROV input was well above the landing base.

Work Team. The rigging team consisted of two rigging "roustabouts," a winch operator, and two supervisors. In retrospect, this was too few riggers. We made some provision for "trouble time," but it is easy to underestimate the amount of rigging that this system requires, especially if problems are encountered. An additional two roustabouts would have been ideal.

In addition to the rigging crew, a dedicated six-man ROV crew and the normal vessel crew were employed. The importance of a highly skilled and motivated ROV crew cannot be overstated. Their experience can optimize the planning and execution of the subsea operations, and a skilled operator can simplify and expedite critical operations such as payload landing and tool operation.

Conclusions

The HCLS has been proven to work under the test conditions, i.e., 3,000 ft water depth, moderate surface currents, moderate Gulf of Mexico seas, and payloads between 2,000-4,000 lbs. It allows the use of a low-cost workboat to land and recover delicate subsea payloads. Care must be taken to follow appropriate rigging practices and avoid system entanglement. Once deployed, the system is very stable and easily maneuvered by an ROV and the workboat. BP Exploration, Shell Offshore, Kerr-McGee, and Oceaneering International are further developing the concept into a working tool to support maintenance operations for their Gulf of Mexico subsea assets.

Acknowledgments

The authors would like to thank their employers and project partners at BP Exploration, BHP Petroleum, Exxon Company USA, Kerr-McGee Corporation, Lazer, Inc., Oceaneering International, Inc., Shell Offshore, and Shell E&P Technology Company.

Nomenclature

HCLS =	Heave Compensated Landing System
ROV =	Remotely Operated Vehicle
6x25 EIPS =	6 Strand Core surrounded by 25 Strand Enhanced Improved Plow Steel

References

1. Langer, C.G., and Kopp, F., U.S. Patent No. 5,190,170.
2. Vogel, Michael (1995), ***Private Communication***, Shell E&P Technology Co., 19951.
3. Morrison, Denby G., Swanson, Richard C., and Pritchard, John J. (1996), **"Dynamics of a Novel Heave-Compensated Deepwater Intervention System,"** ***Proceedings,*** 15th International Conference on Offshore Mechanics and Arctic Engineering, Florence, Italy, June, 1996.
4. Flexcom 3D., (1995), **"Three Dimensional Nonlinear Time Domain Offshore Analysis Software,"** ***Documentation,*** Version 3.1, MCS International, Galway, Ireland.
5. Kopp, F. and Phifer, E.H., (1989), **"Deepwater Pipeline Repair Methods,"** ***Internal Communication,*** Shell Oil Company, Houston, Texas, Dec. 1989.
6. Kopp, F., and Jefferies, R.K., (1989), **"Diverless Buckle Recovery System for Shell's 12" Bullwinkle Pipeline Installation,"** ***Proceedings,*** 8th International Conference on Offshore Mechanics and Arctic Engineering, The Hague, The Netherlands, March 19-23, 1989, pp 77-81.
7. Haws, J.H. and Swanson, Richard C. (1996) , ***Private Communication***, Shell E&P Technology Co., 1996.
8. Swanson, Richard C. (1996), ***Private Communication***, Shell E&P Technology Co., 1996.
9. Shatto, H., (1996), ***Private Communication & Shell Internal Documents***, Shatto Engineering, 1996.

Figures

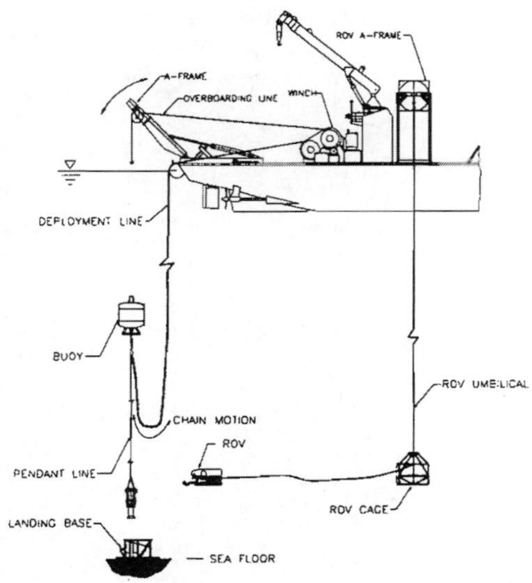

Figure 1- Fully deployed HCLS system.

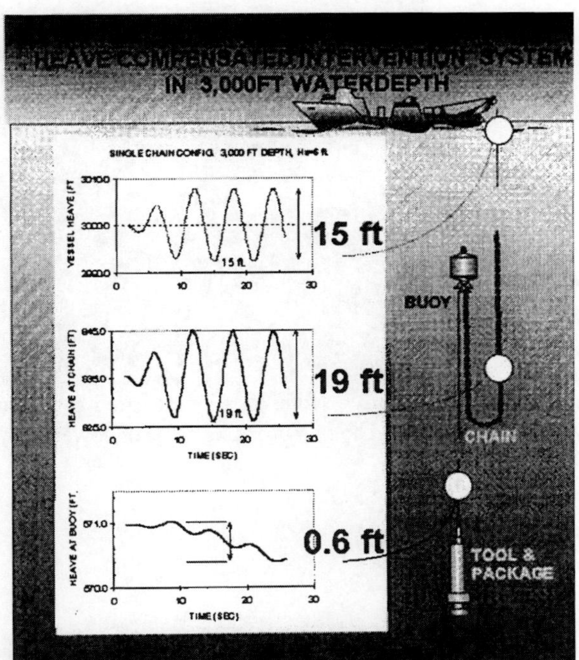

Figure 2- Calculated HCLS system response relative to each other. Note, dampening of subsea equipment relative to vessel stern motion.

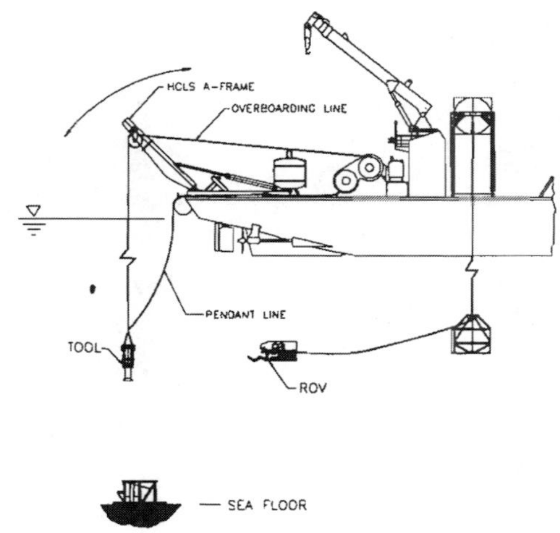

Figure 3-Overboarding running tool with payload package. Note, load carried by overboarding line on a-frame.

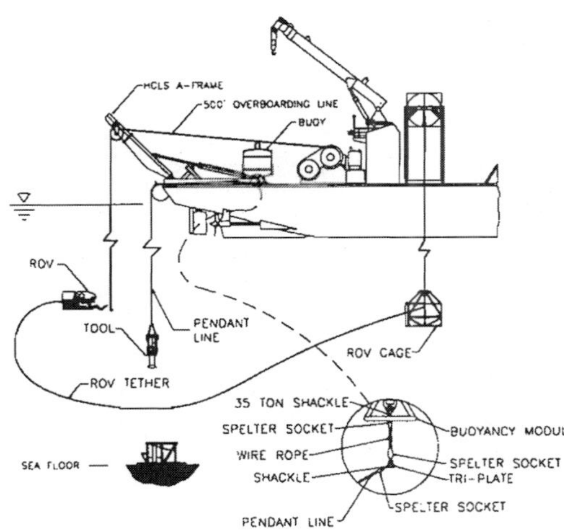

Figure 4- Releasing running tool from overboarding line. Note, load transferred from a-frame to pendant line.

Photo 1- Tool shown in launch position.

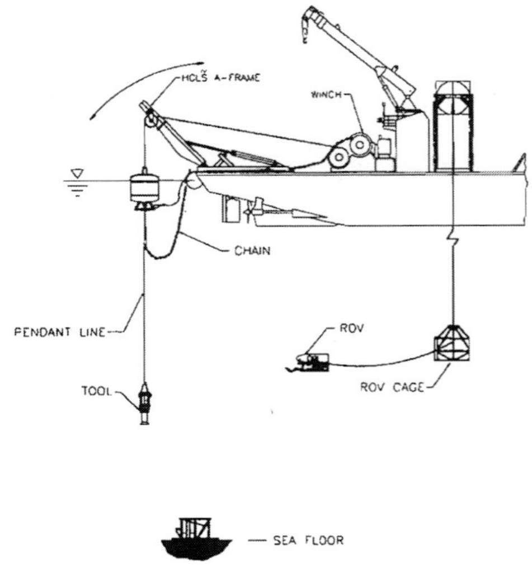

Figure 6- Buoy overboard with tool suspended and chain loaded on winch drum.

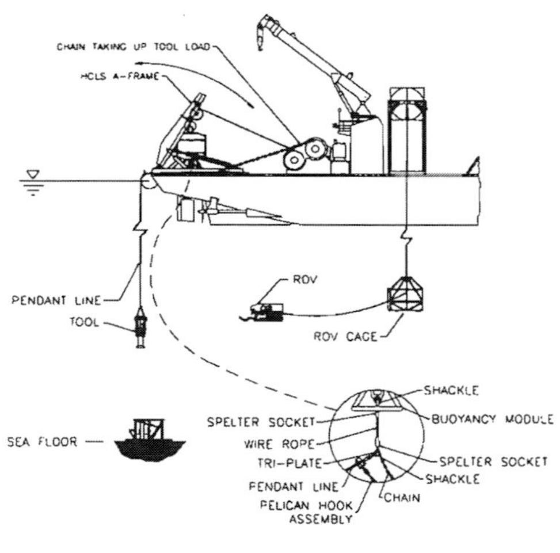

Figure 5- Overboarding buoy with detail showing tri-plate connection. Note, tool load carried by chain on winch drum.

Photo 2- Launching buoy. Note, tool load carried by chain. Buoy currently suspended by overboarding cable on lower drum.

Author Index